Praise for

CONNECTOME

DISCARD

"Sebastian Seung is widely recognized as a rising star in neuroscience. His enthusiasm and communication skills will be put to good use in explaining a fascinating and tremendously exciting new frontier of brain science."

> — **Steven Pinker, Professor of Psychology, Harvard University,**
> **author of** *The Language Instinct* **and** *How the Mind Works*

"Seung has an intelligent, educated, and powerful voice, with a flair for the well-placed metaphor. I've enjoyed it."

> — **Christof Koch,** *Nature*

"We often look to the stars in excitement and wonder. But the reality is that one of the largest mysteries of the universe is right between our ears. The billions of cells in our brains and the vast network of connections between them tells an amazing story that we are just starting to understand, and they make us who we are. It's an amazing journey and having a smart, knowledgeable, (not to mention handsome) personal guide makes it all the most exciting and fun."

> — **Dan Ariely, James B. Duke Professor of Behavioral Economics,**
> **Duke University, author of** *Predictably Irrational*

"A landmark work, generously written. No other researcher has traveled as deeply into the brain forest and eme~~rged to share its secrets.~~"

> — **David Eagleman,** a~~uthor of~~

"An elegant primer on what's known about how the brain is organized and how it grows. . . . [Seung] makes the case that research of the kind he is proposing might solve the puzzle of what is amiss in the brains of people with schizophrenia, Alzheimer's disease, or autism, and might even lead to corrective or preventive treatments for those disorders."
— *Washington Post*

"Sebastian Seung scales the heights of neuroscience and casts his brilliant eye around, describing the landscape of its past and boldly envisioning a future when we may understand our own brains and thus ourselves."
— **Kenneth Blum, Executive Director, Center for Brain Science, Harvard University**

"The author champions the study of 'connectomes,' the unique wiring diagrams of individual brains . . . enjoy Seung's book as a solid primer on brain physiology and the history of brain mapping."
— *Kirkus Reviews*

"Sebastian Seung can do it all. He's widely recognized as a superb physicist, a whiz with computers, and a path-breaking neuroscientist. *Connectome* shows that he's also a terrific writer, as inspiring as he is clear and good humored."
— **Steven Strogatz, Cornell University, author of *Sync: The Emerging Science of Spontaneous Order***

"In *Connectome*, Sebastian Seung reminds us that the human brain has contemplated itself for centuries. This is an important book, full of refreshingly new science and engaging history, about the essential quest to understand ourselves."
— **Philip A. Sharp, MIT, 1993 Nobel Prize in Physiology or Medicine**

Connectome

HOW THE BRAIN'S WIRING
MAKES US WHO WE ARE

Sebastian Seung

MARINER BOOKS
HOUGHTON MIFFLIN HARCOURT
BOSTON • NEW YORK

First Mariner Books edition 2013
Copyright © 2012 by Sebastian Seung

For information about permission to reproduce selections from this book,
write to Permissions, Houghton Mifflin Harcourt Publishing Company,
215 Park Avenue South, New York, New York 10003.

www.hmhbooks.com

Library of Congress Cataloging-in-Publication Data
Seung, Sebastian.
Connectome : how the brain's wiring makes us who we are / Sebastian Seung.
p. cm.
Includes bibliographical references and index.
ISBN 978-0-547-50818-4   ISBN 978-0-547-67859-7 (pbk.)
I. Title. II. Title: How the brain's wiring makes us who we are.
[DNLM: 1. Brain — anatomy & histology. 2. Brain — physiology.
3. Brain — pathology. 4. Cognition — physiology.
5. Nervous System Physiological Phenomena. WL 300]
612.8'2 — dc23
2011028602   612.82

Book design by Brian Moore

Printed in the United States of America
DOC 10 9 8 7 6 5 4 3 2 1

Figure Credits appear on page 335.

To my beloved mother and father,
for creating my genome
and molding my connectome

CONTENTS

Part V: Beyond Humanity

INTRODUCTION

N O ROAD, NO trail can penetrate this forest. The long and delicate branches of its trees lie everywhere, choking space with their exuberant growth. No sunbeam can fly a path tortuous enough to navigate the narrow spaces between these entangled branches. All the trees of this dark forest grew from 100 billion seeds planted together. And, all in one day, every tree is destined to die.

This forest is majestic, but also comic and even tragic. It is all of these things. Indeed, sometimes I think it is everything. Every novel and every symphony, every cruel murder and every act of mercy, every love affair and every quarrel, every joke and every sorrow — all these things come from the forest.

You may be surprised to hear that it fits in a container less than one foot in diameter. And that there are seven billion on this earth. You happen to be the caretaker of one, the forest that lives inside your skull. The trees of which I speak are those special cells called neurons. The mission of neuroscience is to explore their enchanted branches — to tame the jungle of the mind (see Figure 1).

Neuroscientists have eavesdropped on its sounds, the electrical signals inside the brain. They have revealed its fantastic shapes with meticulous drawings and photos of neurons. But from just a few scattered trees, can we hope to comprehend the totality of the forest?

In the seventeenth century, the French philosopher and mathematician Blaise Pascal wrote about the vastness of the universe:

Let man contemplate Nature entire in her full and lofty majesty; let him put far from his sight the lowly objects that surround him; let him regard that blazing light, placed like an eternal lamp to illuminate the world; let the earth appear to him but a point within the vast circuit which that star describes; and let him marvel that this immense circumference is itself but a speck from the viewpoint of the stars that move in the firmament.

Shocked and humbled by these thoughts, he confessed that he was terrified by "the eternal silence of these infinite spaces." Pascal meditated upon outer space, but we need only turn our thoughts inward to feel his dread. Inside every one of our skulls lies an organ so vast in its complexity that it might as well be infinite.

As a neuroscientist myself, I have come to know firsthand Pascal's

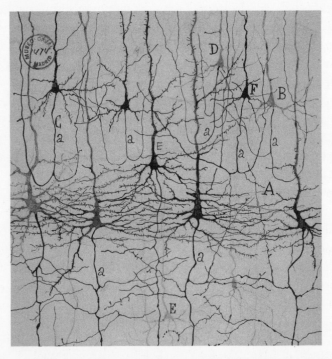

FIGURE 1. *Jungle of the mind: neurons of the cerebral cortex, stained by the method of Camillo Golgi (1843–1926) and drawn by Santiago Ramón y Cajal (1852–1934)*

feeling of dread. I have also experienced embarrassment. Sometimes I speak to the public about the state of our field. After one such talk, I was pummeled with questions. What causes depression and schizophrenia? What is special about the brain of an Einstein or a Beethoven? How can my child learn to read better? As I failed to give satisfying answers, I could see faces fall. In my shame I finally apologized to the audience. "I'm sorry," I said. "You thought I'm a professor because I know the answers. Actually I'm a professor because I know how much I don't know."

Studying an object as complex as the brain may seem almost futile. The brain's billions of neurons resemble trees of many species and come in many fantastic shapes. Only the most determined explorers can hope to capture a glimpse of this forest's interior, and even they see little, and see it poorly. It's no wonder that the brain remains an enigma. My audience was curious about brains that malfunction or excel, but even the humdrum lacks explanation. Every day we recall the past, perceive the present, and imagine the future. How do our brains accomplish these feats? It's safe to say that nobody really knows.

Daunted by the brain's complexity, many neuroscientists have chosen to study animals with drastically fewer neurons than humans. The worm shown in Figure 2 lacks what we'd call a brain. Its neurons are scattered throughout its body rather than centralized in a single organ. Together they form a nervous system containing a mere 300 neurons. That sounds manageable. I'll wager that even Pascal, with his depressive tendencies, would not have dreaded the forest of *C. elegans*. (That's the scientific name for the one-millimeter-long worm.)

Every neuron in this worm has been given a unique name and has a characteristic location and shape. Worms are like precision machines

FIGURE 2. *The roundworm* C. elegans

mass-produced in a factory: Each one has a nervous system built from
the same set of parts, and the parts are always arranged in the same
way.

What's more, this standardized nervous system has been mapped
completely. The result — see Figure 3 — is something like the flight
maps we see in the back pages of airline magazines. The four-letter name of each neuron is like the three-letter code for each of the
world's airports. The lines represent *connections* between neurons,
just as lines on a flight map represent routes between cities. We say
that two neurons are "connected" if there is a small junction, called a
synapse, at a point where the neurons touch. Through the synapse one
neuron sends messages to the other.

Engineers know that a radio is constructed by wiring together
electronic components like resistors, capacitors, and transistors. A
nervous system is likewise an assembly of neurons, "wired" together
by their slender branches. That's why the map shown in Figure 3 was
originally called a wiring diagram. More recently, a new term has been
introduced — *connectome.* This word invokes not electrical engineer-

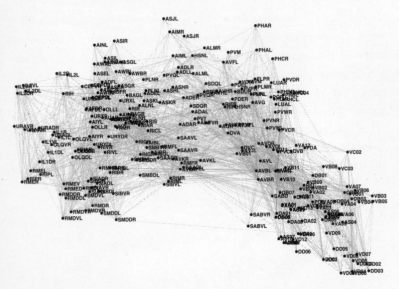

FIGURE 3. *Map of the* C. elegans *nervous system, or "connectome"*

ing but the field of genomics. You have probably heard that DNA is a long molecule resembling a chain. The individual links of the chain are small molecules called nucleotides, which come in four types denoted by the letters A, C, G, and T. Your *genome* is the entire sequence of nucleotides in your DNA, or equivalently a long string of letters drawn from this four-letter alphabet. Figure 4 shows an excerpt from the three billion letters, which would be a million pages long if printed as a book.

```
>gi|224514737|ref|NT_009237.18| Homo sapiens chromosome
11 genomic contig, GRCh37.p5 Primary Assembly
GAATTCTACATTAGAAAAATAAACCATAGCCTCATCACAGGCACTTAAATACACTGAAGCTGCCAAAACA
ATCTATCGTTTTGCCTACGTACTTATCAACTTCCTCATAGCAAACTGGGAGAAAAAAGCAATGGAATGAA
TAAAATGATAGCCACAAAAATCAAGGTGGGAGAAATACTTATTATATGTCCATAAAAAATTTTAATTAAT
GCAAAGTATTAACACCAATGATTGCAGTAATACAGATCTTACAAATGATAGTTTTAGTCTGAACAGGACT
ATCCAAAAGTTAATTTTCTATAGTAACAGTTTTTAAATAAAATATCAATTCCTGAAACACATAAAATGGT
CCATGAGTATACAACGAGTGAAAAAAAACAAATTCAGAGCAAAGATAAATTAAGAAGTATCTAATATTCA
AACATAGTCAAAGAGAGGGAGATTTCTGGATAATCACTTAAGCCCATGGTTAAACATAAATGCAAATATG
TTAATGTTTACTGAATAACTTATCTGTGCCAAGTGGTGTATTAATGATTCATTTTTATTTTTCACTAAAT
CTTTTCTCTAAAGTTGGTGTAGCCTGCAACTAAATGCAAGAAATCTGACCTAGGACCTGCACTTCTTACC
ATTTTGCTCATATTTATTCCCTGTGCATTTTTGTAACATGTATATGTTATATATATAGAAAGAGAGAGAG
GCAGAGATGGAAAGTAATTTATGGAGTTTGATGTTATGTCAGGGTAATTACATGATTATATAATTAACAG
GTTTCTTTTTAAATCAGCTATATCAATAGAAAAATAAATGTAGGAATCAAGAGACTCATTCTGTCCATCT
GTGATAGTTCCATCATGATACTGCATTGTCAAGTCATTGCTCCAAAAATATGGTTTAGCTCAACACTGAG
TGACTATAGGAAACCAGAAACCAGGCTGGGCGCTAAAGATGCAAAGATGAATGAGACATCATCTCTGCCG
TCCAAAAGCTTACTGTCTAGTGGGAGAGTTACACACGTAAGGACAGTAATCTAATAAGAGCTAATAAGTG
AAAACTAAGATAAATTAATAATACAAGATTACAGGGAAGGTTTCCAAAGTCAATGAGGCCTCAAATGAAT
CTTGAAAGTGTGCAAGGATTAACCAAATGAAGAAATGTGTAAGTTTTTCAAACAAAAAGGAACAGCATGA
GCAAATGCAAGGAGGCCTAAAATAAAGAGATGTGTAAAGAGGTGTAAGCAGCTTTGTGCTACTGCCTGAT
AATTAGAAGAATATCGGGAGTAACAAGAGCTATAGAAGAGAGTCACAATTATGGAAAAATATTTATTAAA
TTATAAGAAATTTATAGCATAAGGAATAGTAGGACCATTAAATGTTTTAATAAAGATGATGCTTCTTTTT
TAATATTTATTTTTATTATACTTTAAGTTCTAGGGTACATGTGCACAACGTGCAGGTTACATATGTATAC
ATGTGCCGTGTTGGTGTGCTGCACCCATTAACTCATCATTTACATTAGGTATGTCTCCTAATGCTATCCC
TCCCCCCTCCCCCAACCCCACAACAGGCCGCGGTGTGTGATATTCCCCTTCCTGTGTCCAAGTGTTCTCA
TTGTTCAAGTCCCACCTATGAGTGAAAACATGCGGTGTTTGGTTTTTTGTTCTTGAGATAGATGATGCTT
TAAATTGACCACTCTAGCTGCATTGTGGGAGGAAAAAAAGATTTTAAAACAAGACTAGAAACAGAATAAT
TAGAAAAATGCAACTACAATGCAGATGAGTGATTATCAAGGTCTGAACTGAATAGTGGAAATAGAGATAA
```

FIGURE 4. *A short excerpt from a human genome*

In the same way, a connectome is the totality of connections between the neurons in a nervous system. The term, like *genome*, implies completeness. A connectome is not one connection, or even many. It is *all* of them. In principle, your brain could also be summarized by a diagram that is like the worm's, though much more complex. Would your connectome reveal anything interesting about you?

The first thing it would reveal is that you are unique. You know this, of course, but it has been surprisingly difficult to pinpoint where, precisely, your uniqueness resides. Your connectome and mine are very different. They are not standardized like those of worms. That's consistent with the idea that every human is unique in a way that a worm is not (no offense intended to worms!).

Differences fascinate us. When we ask how the brain works, what mostly interests us is why the brains of people work so differently. Why can't I be more outgoing, like my extroverted friend? Why does my son find reading more difficult than his classmates do? Why is my teenage cousin starting to hear imaginary voices? Why is my mother losing her memory? Why can't my spouse (or I) be more compassionate and understanding?

This book proposes a simple theory: Minds differ because connectomes differ. The theory is implicit in newspaper headlines like "Autistic Brains Are Wired Differently." Personality and IQ might also be explained by connectomes. Perhaps even your memories, the most idiosyncratic aspect of your personal identity, could be encoded in your connectome.

Although this theory has been around a long time, neuroscientists still don't know whether it's true. But clearly the implications are enormous. If it's true, then curing mental disorders is ultimately about repairing connectomes. In fact, any kind of personal change — educating yourself, drinking less, saving your marriage — is about changing your connectome.

But let's consider an alternative theory: Minds differ because genomes differ. In effect, we are who we are because of our genes. The new age of the personal genome is dawning. Soon we will be able to find our own DNA sequences quickly and cheaply. We know that genes play a role in mental disorders and contribute to normal variation in personality and IQ. Why study connectomes if genomics is already so powerful?

The reason is simple: Genes alone cannot explain how your brain got to be the way it is. As you lay nestled in your mother's womb, you already possessed your genome but not yet the memory of your first kiss. Your memories were acquired during your lifetime, not before. Some of you can play the piano; some can ride a bicycle. These are learned abilities rather than instincts programmed by the genes.

Unlike your genome, which is fixed from the moment of conception, your connectome changes throughout life. Neuroscientists have already identified the basic kinds of change. Neurons adjust, or "reweight," their connections by strengthening or weakening them. Neurons reconnect by creating and eliminating synapses, and they rewire by growing and retracting branches. Finally, entirely new neurons are created and existing ones eliminated, through regeneration.

We don't know exactly how life events — your parents' divorce, your fabulous year abroad — change your connectome. But there is good evidence that all four R's — reweighting, reconnection, rewiring, and regeneration — are affected by your experiences. At the same time, the four R's are also guided by genes. Minds are indeed influenced by genes, especially when the brain is "wiring" itself up during infancy and childhood.

Both genes and experiences have shaped your connectome. We must consider both historical influences if we want to explain how your brain got to be the way it is. The connectome theory of mental differences is compatible with the genetic theory, but it is far richer and more complex because it includes the effects of living in the world. The connectome theory is also less deterministic. There is reason to believe that we shape our own connectomes by the actions we take, even by the things we think. Brain wiring may make us who we are, but we play an important role in wiring up our brains.

To restate the theory more simply:

You are more than your genes. You are your connectome.

If this theory is correct, the most important goal of neuroscience is to harness the power of the four R's. We must learn what changes in the connectome are required for us to make the behavioral changes we hope for, and then we must develop the means to bring these changes about. If we succeed, neuroscience will play a profound role in the effort to cure mental disorders, heal brain injuries, and improve ourselves.

Given the complexity of connectomes, however, this challenge is truly formidable. Mapping the *C. elegans* nervous system took over a dozen years, though it contains only 7,000 connections. Your connectome is 100 billion times larger, with a million times more connections than your genome has letters. Genomes are child's play compared with connectomes.

Today our technologies are finally becoming powerful enough that we can take on the challenge. By controlling sophisticated microscopes, our computers can now collect and store huge databases of brain images. They can also help us analyze the torrential flow of data to map the connections between neurons. With the aid of machine intelligence, we will finally see the connectomes that have eluded us for so long.

I am convinced that it will become possible to find human connectomes before the end of the twenty-first century. First we'll move from worms to flies. Later we'll tackle mice, then monkeys. And finally we'll take on the ultimate challenge: an entire human brain. Our descendants will look back on these achievements as nothing less than a scientific revolution.

Do we really have to wait decades before connectomes tell us something about the human brain? Fortunately, no. Our technologies are already powerful enough to see the connections in small chunks of brain, and even this partial knowledge will be useful. In addition, we can learn a great deal from mice and rats, our close evolutionary cousins. Their brains are quite similar to ours and are governed by some of the same principles of operation. Examining their connectomes will shed new light on *our* brains as well as theirs.

In the year A.D. 79, Mount Vesuvius erupted with fury, burying the Roman town of Pompeii under tons of volcanic ash and lava. Frozen in time, Pompeii lay waiting for almost two millennia until it was accidentally rediscovered by construction workers. When archaeologists began to excavate in the eighteenth century, they discovered to their amazement a detailed snapshot of the life of a Roman town — luxurious holiday villas of the wealthy, street fountains and public baths, bars and brothels, a bakery and a market, a gymnasium and a theater, frescoes depicting daily life, and phallic graffiti everywhere. The dead city was a revelation, giving insight into the minutiae of Roman life.

Right now, we can conceive of finding connectomes only by analyzing images of dead brains. You could think of this as brain archaeology, but it's more conventionally known as neuroanatomy. Generations of neuroanatomists have gazed at the cold corpses of neurons in their microscopes and tried to imagine the past. A dead brain, its molecules fastened in place by embalming fluid, is a monument to the thoughts and feelings that once lived inside. Until now, neuroanatomy resembled the act of reconstructing an ancient civilization from the fragmentary evidence of coins and tombs and pottery shards. But connectomes will be detailed snapshots of entire brains, like Pompeii stopped in its tracks. These snapshots will revolutionize the neuroanatomist's ability to reconstruct the functioning of the living brain.

But, you ask, why study dead brains when there are fancy technologies for studying live ones? Wouldn't we learn more if we could travel back in time and study a living Pompeii? Not necessarily. To see why not, imagine some limitations on our ability to observe the living town. Let's say we could watch the actions of a single townsperson but would be blind to all other inhabitants. Or let's say we could look at infrared satellite images revealing the average temperature of each neighborhood but could not see finer details. With such constraints, studying the living town might turn out to be less illuminating than we'd hoped.

Our methods for studying living brains have similar limitations. If we open up the skull, we can see the shapes of individual neurons and measure their electrical signals, but what's revealed is only a tiny fraction of the billions of neurons in the brain. If we use noninvasive imaging methods for penetrating the skull and showing us the brain's interior, we can't see individual neurons; we must settle for coarse information about the shape and activity of brain regions. We can't rule out the possibility that some advanced technology of the future will remove these limitations and enable us to measure the properties of every single neuron inside a living brain, but for now it's just a fantasy.

Measurements of living and dead brains are complementary, and the most powerful approach, in my view, combines them.

Many neuroscientists don't agree with the idea that dead brains can be informative and useful, however. Studying living brains is the only true way of doing neuroscience, they say, because:

You are the activity of your neurons.

Here "activity" refers to the electrical signaling of neurons. Measurements of these signals have provided ample evidence that the neural activity in your brain at any given moment encodes your thoughts, feelings, and perceptions in that instant.

How does the idea that you are the activity of your neurons square with the notion that you are your connectome? Though the two claims might seem contradictory, they are in fact compatible, because they refer to two different notions of the self. One self changes rapidly from moment to moment, becoming angry and then cheering up, thinking about the meaning of life and then the household chores, watching the leaves fall outside and then the football game on television. This self is the one intertwined with consciousness. Its protean nature derives from the rapidly changing patterns of neural activity in the brain.

The other self is much more stable. It retains memories from childhood over an entire lifetime. Its nature — what we think of as personality — is largely constant, a fact that comforts family and friends. The properties of this self are expressed while you are conscious, but they continue to exist during unconscious states like sleep. This self, like the connectome, changes only slowly over time. This is the self invoked by the idea that you are your connectome.

Historically, the conscious self is the one that has attracted the most attention. In the nineteenth century, the American psychologist William James wrote eloquently of the stream of consciousness, the continuous flow of thoughts through the mind. But James failed to note that every stream has a bed. Without this groove in the earth, the water would not know in which direction to flow. Since the connectome defines the pathways along which neural activity can flow, we might regard it as the streambed of consciousness.

The metaphor is a powerful one. Over a long period of time, in the same way that the water of the stream slowly shapes the bed, neural activity changes the connectome. The two notions of the self — as both the fast-moving, ever-changing stream and the more stable but slowly transforming streambed — are thus inextricably linked. This book is about the self as the streambed, the self in the connectome — the self that has been neglected for too long.

In the pages ahead, I will present my vision for a new field of science: connectomics. My primary goal is to imagine the neuroscience of the future and share my excitement about what we'll discover. How can we find connectomes, understand what they mean, and develop new methods of changing them? But we cannot chart the best course forward until we understand where we came from, so I'll start by explaining the past. What do we already know, and where are we stuck?

The brain contains 100 billion neurons, a fact that has overwhelmed even the most fearless explorers. One solution, as I explain in Part I, is to forget about neurons and instead divide the brain into a small number of regions. Neurologists have learned much about the functions of these regions by interpreting the symptoms of brain damage. In developing this method, they were inspired by the nineteenth-century school of thought known as phrenology.

Phrenologists explained mental differences as arising from variations in the *sizes* of the brain and its regions. By imaging the brains of many human subjects, modern researchers have confirmed this idea, using it to explain differences in intelligence as well as mental disorders like autism and schizophrenia. They have found some of the strongest evidence we have for the idea that minds differ because brains differ. The evidence is statistical, however — revealed only by averages over populations. The sizes of the brain and its regions remain almost useless for predicting the mental properties of an individual.

This limitation is no mere technicality. It is fundamental. Although phrenology assigns functions to brain regions, it does not attempt to explain *how* each region performs its function. Without that, we cannot explain in a satisfying way why the region might function especially well in some people and malfunction in others. We can, and must, find a less superficial answer than size.

In Part II, I introduce an alternative to phrenology called *connectionism,* which also dates back to the nineteenth century. This approach is conceptually more ambitious, because it attempts to explain how regions of the brain actually work. Connectionists view a brain region not as an elementary unit but as a complex network composed of a large number of neurons. The connections of the network are organized so that its neurons can collectively generate the intricate patterns of activity that underlie our perceptions and thoughts. The organization of connections can be altered by experience, which allows us to learn and remember. The organization is also shaped by genes, as described in Part III, so that genetic influences on the mind can also be explained. These ideas may sound powerful, but there is a catch: They have never been subjected to conclusive experimental tests. Connectionism, despite its intellectual appeal, has never managed to become real science, because neuroscientists have lacked good techniques for mapping the connections between neurons.

In a nutshell, neuroscience has been saddled with a dilemma: The ideas of phrenology can be empirically tested but are simplistic. Connectionism is far more sophisticated, but its ideas cannot be evaluated experimentally. How do we break out of this impasse? The answer is to find connectomes and learn how to use them.

In Part IV, I explore how this will be done. We are already starting to develop technologies for finding connectomes, and I'll describe the cutting-edge machines that will soon be hard at work in labs around the world. Once we find connectomes, what will we do with them? First, we'll use them to carve the brain into regions, aiding the work of neo-phrenologists. And we'll divide the enormous number of neurons into types, much as botanists classify trees into species. This will dovetail with the genomic approach to neuroscience, because genes exert much of their influence on the brain by controlling how neuron types wire up with each other.

Connectomes are like vast books written in letters that we barely see, in a language that we do not yet comprehend. Once our technologies make the writing visible, the next challenge will be to understand what it means. We'll learn to decode what is written in connectomes by attempting to read memories from them. This endeavor will at long last provide a conclusive test of connectionist theories.

But it won't be enough to find a single connectome. We will want to find many connectomes and compare them, to understand why one mind differs from another, and why a single mind changes over time. We'll hunt for *connectopathies,* abnormal patterns of neural connectivity that might underlie mental disorders such as autism and schizophrenia. And we'll look for the effects of learning on connectomes.

Armed with this knowledge, we will develop new methods of changing connectomes. The most effective way at present is the traditional one: training our behaviors and thoughts. But learning regimens will become more powerful when supplemented by molecular interventions that promote the four R's of connectome change.

The new science of connectomics will not be established overnight. Today we can only see the beginning of the road, and the many barriers that lie in the way. Nevertheless, over the coming decades, the march of our technologies and the understanding that they enable will be inexorable.

Connectomes will come to dominate our thinking about what it means to be human, so Part V concludes by taking the science to its logical extreme. The movement known as transhumanism has developed elaborate schemes for transcending the human condition, but are the odds in their favor? Does the ambition of cryonics to freeze the dead and eventually resurrect them have any chance of succeeding? And what about the ultimate cyber-fantasy of uploading, of living happily ever after as a computer simulation, unencumbered by a body or a brain? I will attempt to extract some concrete scientific claims from these hopes and propose how to test them empirically using connec-tomics.

But let's not entertain such heady thoughts about the afterlife just yet. Let's begin by thinking about this life. In particular, let's start with the question mentioned earlier, the one that everyone has thought about at some point: Why are people different?

PART I

DOES SIZE MATTER?

ONE

Genius and Madness

I N 1924 ANATOLE FRANCE died near Tours, a city on the Loire River. While the French nation mourned their celebrated writer, anatomists from the local medical college examined his brain and found that it weighed merely 1 kilogram, about 25 percent less than average. His admirers were crestfallen, but I don't think they should have been surprised. In the photographs of Figure 5, Anatole France looks like a pinhead next to the Russian writer Ivan Turgenev.

Sir Arthur Keith, one of the most prominent anthropologists in England, expressed his perplexity:

> Although we know nothing of the finer structural organization of Anatole France's brain, we do know that with it he was performing feats of genius while millions of his fellow countrymen, with brains 25 percent or even 50 percent larger, were manifesting the average abilities of daily labourers.

Anatole France was a "man of average size," Keith noted, so the smallness of his brain could not be explained away by invoking a small body. Keith went on to express his bemusement:

Ivan Turgenev, 1818–1883   Anatole France, 1844–1924
2,021 grams   1,017 grams

FIGURE 5. *Two famous writers whose brains were examined and weighed after death*

This lack of correspondence between brain mass and mental ability
. . . has been a lifelong puzzle to me. I have known . . . men with the
most massive heads and sagacious appearances who proved failures
in all the trials to which the world submitted them, and I have known
small-headed men succeed brilliantly, just as Anatole France did.

Keith's confession of ignorance surprised me with its honesty, and
the thought of Anatole France as a neural David triumphing over a
world of Goliaths made me chuckle. At a scientific seminar I once read
Keith's words out loud. A French theoretical physicist shook his head
and commented wryly, "Anatole France was not such a great writer after all." The audience laughed, and laughed again when I noted that his
amateur scribbles had earned him the 1921 Nobel Prize in Literature.

The case of Anatole France shows that brain size and intelligence are
unrelated for individuals. In other words, you cannot use one to reliably predict the other for any given person. But it turns out that the
two quantities have a *statistical* relationship — one that's revealed by
averages over large populations of people. In 1888 the English poly-

math Francis Galton published a paper entitled "On Head Growth in Students at the University of Cambridge." He divided students into three categories based on their grades, and showed that the average head size of the best students was slightly larger than that of the worst students.

Many variations on Galton's study have been done over the years, using methods that have become more sophisticated. School grades were replaced by standardized tests of intellectual abilities, colloquially known as IQ tests. Galton estimated head volume by measuring length, width, and height and then multiplying the numbers. Other investigators measured head circumference using a tape. The most intrepid preferred to remove and weigh the brains of the deceased. All of these methods seem primitive, now that researchers can see the living brain right through the skull using magnetic resonance imaging (MRI). This amazing technology generates cross-sectional images of the brain like the one shown in Figure 6.

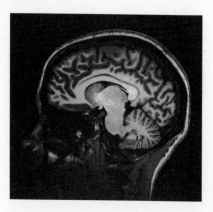

FIGURE 6. *An MRI cross-section of the brain*

In effect, MRI virtually cuts the head into slices and generates a two-dimensional (2D) image of each slice. From the resulting "stack" of 2D images, researchers can reconstruct the entire shape of the brain in three dimensions (3D) and then calculate the volume of the brain very accurately. Because of MRI, it has become much easier to conduct studies relating IQ to brain volume. From many studies of this kind over the past two decades, the consensus is clear: On average, people with bigger brains have higher IQs. Modern studies with improved methods have confirmed Galton.

This confirmation, however, does not contradict what we learned from Anatole France. Brain size is still almost useless for predicting the IQ of an individual person. What exactly do I mean by "almost useless"? If two variables are statistically related, they are said to be *correlated*. Statisticians grade the strength of any correlation with a single number known as Pearson's correlation coefficient, which ranges between the limits -1 and $+1$. If this number — usually designated by the letter r — is close to the limits, the correlation is strong, meaning that if you know one variable, you can predict the other with high accuracy. If r is close to zero, the correlation is weak; you will be highly inaccurate if you attempt to use one variable to predict the other. The correlation between IQ and brain volume is about $r = 0.33$, which is quite weak.

The moral of the story is that statistical statements about averages should not be interpreted as being about individual persons. The misinterpretation is easy to make and easy to foster, which is one reason for the quip that there are three kinds of lies: lies, damned lies, and statistics.

The scientific papers in this line of research are dignified by scholarly language, not to mention loads of footnotes and citations, but one can't escape the feeling that all this measuring of heads is kind of funny. Indeed, Galton the man was kind of funny — as in peculiar. His motto, "Whenever you can, count," captures his obsessive love of quantification, which bordered on the ludicrous. In his memoirs he recounted an attempt to create a "Beauty Map" of Britain. While walking the streets of a city, he would prick holes in a piece of paper he held surreptitiously in his pocket. The holes recorded the beauty of the women he passed, ranked as "attractive," "indifferent," or "repellent." The result of his study? "I found London to rank highest for beauty; Aberdeen lowest."

There is also an insulting aspect to this line of research. The famous statistician Karl Pearson, Galton's protégé and the inventor of the correlation coefficient, ordered people on a linear scale with nine divisions: genius, specially able, capable, fair intelligence, slow intelligence, slow, slow dull, very dull, and imbecile. Summarizing a person by a single number or category — whether the summary is of intelligence, beauty, or any other personal characteristic — is reductionist and dehumanizing. Some researchers have crossed the line from insulting to immoral, using their studies to advocate extreme policies of eugenics and racial discrimination.

Yet it would be a mistake to simply reject Galton's finding because it seems silly, or because it can be misused, or because the correlation is weak. On the positive side, Galton provided the basis for a plausible hypothesis: Differences in the mind are caused by differences in the brain. He used the best method available to him, looking at the relationship between grades in school and head size. Contemporary researchers use IQ and brain size, measures that are better but still crude. If we continued to refine our measures, might we discover correlations that are much stronger?

Summarizing the brain's structure by a single number like total volume or weight seems superficial. Even casual examination of the brain reveals multiple regions, each of which looks very different to the naked eye. The cerebrum, the cerebellum, and the brainstem — shown in Figure 7 — are plainly visible when the brain is removed intact from the skull, as was done at the autopsies of Anatole France and Ivan Turgenev.

You can imagine the brainstem holding the cerebrum up like fruit on a stalk, with the cerebellum decorating the junction like a leaf. The cerebellum is important for graceful movement, but its removal mostly spares mental abilities. Damage to the brainstem can kill, because it controls many vital functions, such as breathing. Extensive damage to the cerebrum leaves the victim alive but unconscious. The cerebrum is widely regarded as the most important of the three parts

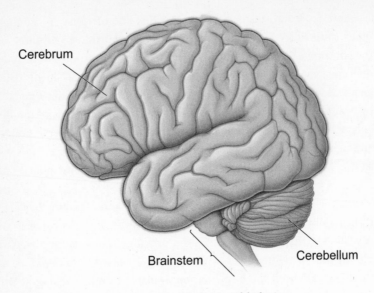

Cerebrum

Brainstem

Cerebellum

FIGURE 7. *A tripartite division of the brain*

for human intelligence; it is critical for virtually all our mental abilities. It is also the largest of the three parts, occupying about 85 percent of total brain volume.

Most of the surface of the cerebrum is covered by a sheet of tissue just a few millimeters thick. This is known as the *cerebral cortex,* or cortex for short. Spanning the area of a hand towel, the cortex can fit inside the skull only because it's folded up. The folds give the cerebrum a wrinkled appearance. The most obvious boundary within the cortex is visible from above: a large groove running from front to back (see Figure 8, left). This groove, called the longitudinal fissure, divides the left and right hemispheres of the cerebrum, the "left brain" and "right brain" of pop psychology.

It's less obvious how to subdivide each cerebral hemisphere, but one reasonable approach relies again on grooves of the cortex. After the longitudinal fissure, the next most prominent groove is called the Sylvian fissure (see Figure 8, right). After that is the central sulcus, which runs vertically from the Sylvian fissure toward the top of the brain. These two major grooves divide each hemisphere into four lobes:

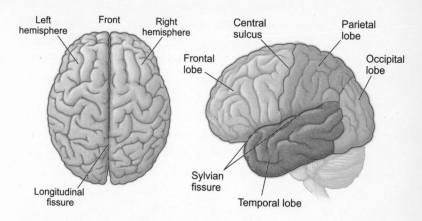

FIGURE 8. *The cerebrum divided into hemispheres* (left), *and each hemisphere divided into lobes* (right)

frontal, parietal, occipital, and temporal. (By the way, it's worth memorizing the names and locations of these lobes, as I will refer to them often.)

There are many other, more minor grooves on the brain's surface, some of which are in roughly the same location from person to person. These have names and are still used today as landmarks. But does dividing the cortex along its grooves really make sense? Are they genuine boundaries, or merely an insignificant byproduct of the fact that the cortex has to fold to fit inside the skull?

The problem of dividing the cortex was first confronted in the nineteenth century. Before then, it was thought that the cortex merely served to cover the rest of the brain. (The term *cortex* is derived from the Latin word meaning "bark," as in tree bark.) In 1819 the German physician Franz Joseph Gall published his theory of "organology." He noted that every organ of the body serves a distinct function: the stomach for digestion, the lungs for breathing, and so on. Gall argued that the brain is too complex to be a single organ, and the mind too complex to be a single function. He proposed to divide both. In particular, he recognized the importance of the cortex and divided it into a set of regions, which he called the "organs" of the mind.

Gall's disciple Johann Spurzheim later introduced the term *phrenology,* more familiar to us than Gall's original name for the theory. The phrenological map shown in Figure 9 displays regions corresponding to functions with names like "acquisitiveness," "firmness," and "ideality." These particular correspondences are now considered fanciful imaginings based on flimsy evidence, but the phrenologists eventually turned out to be more right than wrong. Their emphasis on the cortex is widely accepted today, and their approach of localizing mental functions to particular cortical regions is still taken seriously. It now goes by the name of cortical or cerebral *localizationism.*

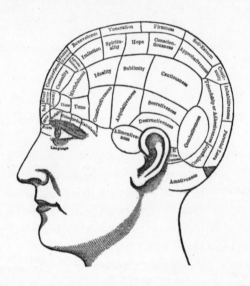

FIGURE 9. *Phrenological map*

The first real evidence for localization came later in the nineteenth century from observations of patients with brain damage. At that time, many French neurologists worked at two Parisian hospitals. Salpêtrière, on the Left Bank of the Seine River, housed female patients; male patients were placed farther from the city center, in Bicêtre. Both hospitals were founded in the seventeenth century and had functioned as prisons and mental asylums too. (The distinction was blurred by Bicêtre's most famous inmate, the Marquis de Sade.)

Both hospitals had pioneered humane methods for the treatment of the insane, such as not confining them in chains. I imagine that they remained depressing places all the same.

In 1861 the French physician Paul Broca was called to examine a fifty-one-year-old patient suffering from an infection in his surgical ward at Bicêtre. According to the records, the patient had been incarcerated since the age of thirty. At the time of admission he had already lost the ability to speak any word except the monosyllable "tan," which became his nickname. Since Tan could communicate with hand gestures, it seemed that he could comprehend language, although he could not speak it.

A few days after the examination, Tan succumbed to his infection, and Broca performed an autopsy. He sawed open the skull, removed the brain, and placed it in alcohol for preservation. The most prominent damage to Tan's brain — see Figure 10 — was a large cavity in the left frontal lobe.

Broca announced his discovery to the Anthropological Society the next day. He claimed that the damaged region in Tan's brain was the source of speech, which was distinct from comprehension. Today, loss of language ability is known as *aphasia*. Loss of speech, in particular, is called Broca's aphasia, and the damaged location in Tan's cerebral cortex is known as Broca's region. With his discovery, Broca man-

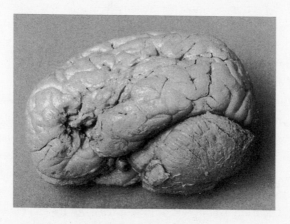

FIGURE 10. *Tan's brain, with damage to Broca's region*

aged to settle a debate that had raged for decades. The phrenologist Gall had asserted at the beginning of the nineteenth century that linguistic functions were located in the frontal lobe of the brain, but had been met with skepticism. Broca finally provided some convincing evidence, as well as a specific location in the frontal lobe.

As time went on, Broca encountered more cases similar to Tan's and found that they all involved damage to the left hemisphere of the brain. Given that the two hemispheres looked so similar to each other, it was hard to believe that they could be different in their functions. But the evidence mounted, and Broca concluded in an 1865 paper that the left hemisphere was specialized or dominant for language. Subsequent researchers have confirmed that this is the case for almost all people. Thus Broca's findings supported not only cortical localization but also *cerebral lateralization*, the idea that mental functions are located in either the left or the right hemisphere.

In 1874 the German neurologist Carl Wernicke described a different kind of aphasia. Unlike Tan, his patient could speak words fluently, but the sentences didn't make sense. Furthermore, the patient could not comprehend questions asked of him. Autopsy showed damage to part of the temporal lobe of the left hemisphere. Wernicke concluded that loss of comprehension was the primary effect of damage to this region. Production of nonsensical sentences was a secondary effect, which could have arisen because a person may need to comprehend what he or she is saying in order to say something that makes sense. The symptoms caused by damage to Wernicke's region are known today as Wernicke's aphasia.

Together, Broca and Wernicke provided a *double dissociation* of speech production and comprehension. Damage to Broca's region halted production of words but left comprehension intact; damage to Wernicke's region destroyed comprehension but spared production. This was important evidence that the mind is "modular." It might seem obvious that language is distinct from other mental abilities, since it is possessed by humans but not other animals, but it's less obvious — or *was* less obvious, before Broca and Wernicke — that language can be further subdivided into separate modules for production and comprehension.

Broca and Wernicke showed how to map the cortex by relating the symptoms of patients to the locations of brain lesions. By using this method, their successors were able to identify the functions of many other regions of the cortex. They created maps resembling those of the phrenologists, but based on solid data. Could their findings on cortical localization be used to understand mental differences?

When Albert Einstein died in 1955, his body was cremated. His brain was not, because it had been removed by the pathologist Thomas Harvey during an autopsy. Fired from Princeton Hospital a few months later, Harvey kept Einstein's brain. Over the following decades he carried 240 pieces with him in a jar as he moved from city to city. In the 1980s and 1990s, Harvey sent specimens to several researchers who shared his goal of finding out what was special about the brain of a genius.

Harvey had already determined that the weight of Einstein's brain was average, or even slightly below average; thus brain size couldn't explain why Einstein was extraordinary. Sandra Witelson and her collaborators proposed another explanation in 1999. They argued, based on the photographs Harvey had taken during the autopsy, that a cortical region called the inferior parietal lobule was enlarged. (This region is part of the parietal lobe of the brain.) Perhaps Einstein was a genius because *part* of his brain was enlarged. Einstein himself reported that he often thought in images rather than words, and the parietal lobe of the brain is known to be involved in visual and spatial thinking.

Anatole France and Albert Einstein belong to a long tradition of public fascination with geniuses' brains. Nineteenth-century enthusiasts preserved the brains of luminaries like the poets Lord Byron and Walt Whitman, which still sit today in dusty jars relegated to the back rooms of museums. I find it strangely heartening that Tan and Paul

Broca, the wordless patient and the neurologist who studied him, are now companions for eternity, as the same Parisian museum preserves both of their brains. Neuroanatomists also preserved the brain of Carl Gauss, one of the greatest mathematicians of all time. They pointed to an enlarged parietal lobe to explain his genius, anticipating Witelson's explanation of Einstein's.

So the strategy of studying the sizes of specific brain regions rather than overall brain size is not new at all. In fact, it was originally invented by the phrenologists. Their founding father, Franz Joseph Gall, titled his 1819 treatise *The Anatomy and Physiology of the Nervous System in General, and of the Brain in Particular, with Observations upon the possibility of ascertaining the several Intellectual and Moral Dispositions of Man and Animal, by the configuration of their Heads.* Gall held that each mental "disposition" is correlated with the size of the corresponding cortical region. More dubiously, Gall argued that the shape of the skull reflected the shape of the underlying cortex and could be used to divine a person's dispositions. Phrenologists roamed the world offering to predict the fortunes of children, assess prospective marriage partners, and screen job applicants by feeling bumps on heads.

Gall and his disciple Spurzheim proposed functions for cortical regions based on anecdotes about extreme dispositions. If a genius had a large forehead, intelligence must be in the front of the brain. If a criminal's head bulged on the sides, the temporal lobe must be important for telling lies. Their anecdotal methods led to localizations that were mostly preposterous. By the second half of the nineteenth century, phrenology had become an object of ridicule.

Today we have technologies that the phrenologists could only fantasize about. MRI gives us precise measurements of the sizes of cortical regions, eliminating the silly method of feeling head bumps. And by scanning the brains of many humans, researchers can collect enough data to go beyond anecdotes like Witelson's study of Einstein's brain. What have the neo-phrenologists found?

They have demonstrated that IQ is correlated with the sizes of the frontal and parietal lobes. The correlation has turned out to be slightly stronger than that between IQ and overall brain size, in keeping with the idea that these lobes are more critical to intelligence. (The occipital and temporal lobes are mainly devoted to sensory abilities like vision and hearing.) Still, the correlation is disappointingly weak.

But these studies don't fully follow the spirit of phrenology, which not only divided the brain into regions but also divided the mind into separate abilities. We all know people who are superb at mathematics but less skilled verbally, and vice versa. Today many researchers reject the notion of IQ and general intelligence as simplistic. They prefer to speak of "multiple intelligences," and these turn out to be correlated with the sizes of specific brain regions. London taxi drivers have an enlarged right posterior hippocampus, which is a region of the cortex thought to be involved in navigation. In musicians, the cerebellum is larger and certain cortical regions are thicker. (The enlargement of the cerebellum makes sense, as it is thought to be important for fine motor skills.) Bilinguals have a thicker cortex in the lower part of the left parietal lobe.

While these findings are fascinating, they are only statistical. If you read the fine print, you'll see that the brain regions are only larger *on average.* It remains the case that the sizes of brain regions are almost useless for predicting the abilities of an individual.

Differences in intellectual ability can cause difficulties, but they're usually not catastrophic. Other kinds of mental variation, however, exact terrible suffering and are hugely costly to our society. In industrialized countries, an estimated six of every hundred people have a severe mental disorder, and almost half suffer a milder disorder at some point in their lives. Most disorders respond only partially to behavioral and drug therapies, and many have no known treatment at all. Why is it so difficult to fight mental disorders?

The discoverer of a disease is usually the first to describe its symptoms. In 1530 the Italian physician Girolamo Fracastoro utilized the unusual medium of an epic poem, *Syphilis sive morbus Gallicus* ("Syphilis or the French Disease"). He named the disease in honor of the first man to contract it, the mythical shepherd Syphilus, who was punished with sickness by the god Apollo. In three books of Latin hexameter, Fracastoro described the symptoms of syphilis, recognized that it was sexually transmitted, and prescribed some remedies.

Syphilis causes ugly skin lesions and awful physical deformities. Later on, another horrible symptom can emerge: insanity. In his 1887 horror story "Le Horla," the French writer Guy de Maupassant imagined a supernatural being who torments the narrator, first by physical sickness and then by madness: "I am lost! Somebody possesses my soul and governs it! Somebody orders all my acts, all my movements, all my thoughts. I am no longer anything in myself, nothing except an enslaved and terrified spectator of all the things which I do." The narrator finally resolves to end his suffering by killing himself. The story seems semi-autobiographical, as Maupassant suffered from syphilis contracted in his twenties. In 1892 he attempted suicide by cutting his throat. Committed to an asylum, Maupassant died the next year at the age of forty-two.

The painter Paul Gauguin and the poet Charles Baudelaire may also have suffered from syphilis. We have no proof, however, because a disease cannot be reliably diagnosed based on symptoms alone. Two people with the same disease may have different symptoms, and two people with different diseases may have similar symptoms. To diagnose and treat a disease, we'd like to know its cause rather than its symptoms. The bacterial cause of syphilis was discovered in 1905, and the first drugs that killed the bacteria soon followed. These drugs were effective in the early stages of syphilis, but they could not eradicate the disease after it invaded the nervous system. In 1927 the German physician Julius Wagner-Jauregg won the Nobel Prize for his bizarre cure for neurosyphilis. In addition to administering drugs, he deliberately infected patients with malaria. The resulting fever somehow killed off the syphilis bacteria, at which point he introduced drugs that cured the malaria. After World War II, Wagner-Jauregg's cure was replaced by penicillin and the other antibacterial drugs known as antibiotics. Syphilis is no longer a major cause of brain disease.

Diseases caused by infection are relatively easy to cure, because we know the cause. But what about other kinds? Alzheimer's disease (AD), which commonly strikes the elderly, starts with memory loss and progresses to dementia, a generalized deterioration of mental abilities. In the late stages, the brain shrinks, leaving empty space inside the skull. Were they alive today, the phrenologists would explain AD as being caused by a decrease in brain size, but this explanation would be unsatisfactory. Shrinkage of the brain occurs long after memory loss and other symptoms first appear, and furthermore, shrinkage is itself more a symptom than a cause. It happens because brain tissue dies, but what causes that?

Searching for clues, scientists examined autopsy tissue from AD patients and discovered microscopic "junk" called plaques and tangles littering the brains. In general, an abnormality in the cells of the brain associated with a disease is known as a *neuropathology*. Plaques and tangles appear in the brain well before the death of cells, and closer to the onset of AD symptoms. These neuropathologies are currently regarded as the defining characteristic of AD, as the symptoms of memory loss and dementia can also occur in other diseases. Scientists have not yet figured out what causes plaques and tangles to accumulate, but they hope that reducing these neuropathologies might cure AD.

The most puzzling mental disorders come with no clear and consistent neuropathology. Here we are really stumped. These disorders, still defined only by their psychological symptoms, are the furthest from being cured. They may involve anxiety, as in panic and obsessive-compulsive disorders, or mood, as in depression and bipolar disorder. Two of the most debilitating are schizophrenia and autism.

The symptoms of autism are most memorably conveyed by clinical description:

David was 3 when he was diagnosed as autistic. At that time he hardly looked at people, was not talking, and seemed lost in his own world. He loved to bounce on a trampoline for hours and was extremely adept at doing jigsaw puzzles. At 10 years of age David had developed well physically, but emotionally remained very immature. He had a beautiful face with delicate features. . . . He was and still is extremely stubborn in his likes and dislikes. . . . More often than not his mother has to give in to his urgent and repeated demands, which easily escalate into tantrums.

David learned to talk when he was 5. He now goes to a special school for autistic children, where he is happy. He has a daily routine, which he never varies. . . . Some things he learns with great skill and speed. For example, he learned to read all by himself. He now reads fluently, but he doesn't understand what he reads. He also loves to do sums. However, he has been extremely slow to learn other skills, for example, eating at the family table, or getting dressed. . . .

David is now 12 years old. He still does not spontaneously play with other children. He has obvious difficulties in communicating with other people who don't know him well. . . . He makes no concessions to their wishes or interests and cannot take onboard another person's point of view. In this way David is indifferent to the social world and continues to live in a world of his own.

This case study includes all three of the symptoms that define autism: social impairment, difficulties with language, and repetitive or rigid behaviors. The symptoms appear before three years of age and often lessen later on, but most autistic adults are unable to function without some sort of supervision. No known treatment is very effective, and there is certainly no cure.

Speaking more poetically, Uta Frith has described autism as a "beautiful child imprisoned in a glass shell." Many other types of disabled children may have heart-wrenchingly obvious physical deformities. That's not the case for the autistic, who look superficially fine or even beautiful. Their appearance deceives parents, who have difficulty believing that something is fundamentally wrong. They hope in vain to break through the "glass shell" — the social isolation of autism — and liberate a normal child. But the healthy guise of the autistic child hides a brain that is not normal.

The best-documented abnormality is one of size. When the American psychiatrist Leo Kanner originally defined the syndrome in a land-mark 1943 paper, he noted in passing that five children out of his eleven case studies had large heads. Over the years, researchers have studied many more autistic children and found that their heads and brains are indeed enlarged on average — especially the frontal lobe, which contains many areas involved in social and linguistic behaviors.

Does that mean brain size is a good predictor of autism? If it were, we could be confident that the phrenological approach is on the right track toward explaining autism. But we should be careful not to commit a common statistical fallacy concerning rare categories. Consider a very special type of person, professional football players. They are markedly larger than the average person. Can we turn this around and predict that anyone much larger than average is likely to be a professional football player? This prediction rule would work well with what's called a balanced population — one containing equal numbers of football players and regular people. If you sorted them by size, you'd be pretty accurate. But if you looked instead at the general population and predicted that any large person drawn from it was a football player, you'd be wrong most of the time. These people would just be tall, muscular, or obese for other reasons. Similarly, predicting that all children with big brains are autistic would be highly unreliable. There is much more to playing in the NFL than being large, and much more to being autistic than having a big brain.

The media often report studies claiming accurate prediction of rare mental disorders based on some property of the brain. These studies usually turn out to be less impressive than they sound, because the accuracy is only for a balanced population, not for the general population. If, however, you really know a disease's cause, it should serve as an unerring diagnostic, even for the general population. That's the case for many infectious diseases, which can be detected through blood tests for microbes.

Schizophrenia is as perplexing as autism. It typically begins in the twenties, with the striking and sudden onset of hallucinations (most commonly hearing voices), delusions (often of persecution), and dis-

organized thinking. Here is a vivid first-person account of such symptoms, collectively known as psychosis:

> Though I cannot remember how it was initiated, at one point while
> I was sitting on the toilet, a quick rush of adrenaline gripped me.
> My heart was racing. Voices started coming out of nowhere, and I
> thought that I was mentally tuned into a television program being
> broadcast worldwide in which rock stars and scientists were over-
> throwing the world governments (through the means of computers,
> biology, psychology, and voodoo-type ritual). Right then and there!
>
> At that moment the people communicating on TV were an-
> nouncing all of their intentions and motives for a new world order.
> I seemed to be at center stage of the discussion with a number of
> rock stars and scientists who were hiding elsewhere throughout the
> world.

Psychosis can terrify the victim, as well as alarm and distress others. It's the most obvious sign of schizophrenia, but it also accompanies other mental disorders. So an accurate diagnosis of schizophrenia
requires additional symptoms, such as lack of motivation, flattened
emotion, and diminished speech. These are the "negative" symptoms
of schizophrenia, in contrast to the "positive" or psychotic symptoms.
(Here, "positive" and "negative" are not value judgments; they refer to
the presence of disordered thought and the relative absence of emotion, respectively.) Schizophrenia is treated with drugs that eliminate
psychosis. The drugs are not a complete cure, however, because they
are less effective for the negative symptoms. Most schizophrenics remain unable to live independently.

As with autism, the best-documented abnormality of the schizophrenic brain has to do with size. MRI studies have shown that overall brain volume is reduced on average by just a few percentage points.
The percent reduction of the hippocampus is slightly greater but still
not very large. Researchers have also imaged the ventricular system,
a set of fluid-filled caverns and passages in the brain. The lateral and
third ventricles are enlarged on average by 20 percent. Since the ven-

tricles are hollow spaces in the brain, their enlargement might be re-lated to the observed reduction in brain volume. While it's encouraging that some sort of difference has been found, this correlation is as weak as the statistical findings reported for autism. Diagnosing schizophrenia for an individual using brain size, hippocampal size, or ventricular volume would be wildly inaccurate.

To make progress in treating autism and schizophrenia, it would help to find clear and consistent neuropathologies like the plaques and tangles of Alzheimer's disease, but no similar accumulation of "junk," or other signs of dying or degenerating cells, is consistently associated with autistic or schizophrenic brains. Neo-phrenology suggests that something is abnormal about the brain, but we have failed to find it. In 1972 the neurologist Fred Plum wrote despairingly that "schizophrenia is the graveyard of neuropathologists." Researchers have discovered some clues since then, but there has been no dramatic breakthrough.

Most of us are convinced that minds differ because brains differ; so far, however, there is little proof. The phrenologists tried to find evidence by examining the sizes of the brain and its regions, but only recently has MRI provided the technological means to execute their strategy properly. Neo-phrenology has confirmed that mental differences are statistically related to brain size, by revealing weak correlations in groups of people, but the differences do not accurately predict genius, autism, or schizophrenia in individuals.

I wish neuroscience were winning its game more convincingly. The stakes are high. Discovering neuropathologies for autism and schizophrenia could aid the search for therapies. Understanding what makes a brain intelligent could help us devise better teaching methods or other tools to make people smarter. We don't just want to understand the brain. We want to change it.

TWO

Border Disputes

God, grant me the serenity
To accept the things I cannot change
Courage to change the things I can
And wisdom to know the difference.

THE SERENITY PRAYER has been adopted by Alcoholics Anonymous and other organizations that help members recover from addiction. It reveals why the brain fascinates people so much: They are always hoping to change it. Just stroll through the self-help section of your local bookstore — you'll see hundreds of titles on how to drink less, quit drugs, eat right, manage money, discipline your kids, and save your marriage. All these things seem possible, but they are difficult to achieve.

Certainly normal, healthy adults would like to change their behaviors, but this goal is even more critical for those with mental disabilities and disorders. Can a young adult ever be cured of schizophrenia? Can a grandparent learn to speak again after a stroke? And we all want our schools and our childrearing to mold young minds for the better. Can we improve the way this is done?

The Serenity Prayer asks for courage and wisdom about change. Wouldn't it be better to have answers from neuroscience as well? After all, changing the mind is ultimately about changing the brain. But neuroscience can never aid the quest for self-improvement without answering a more fundamental question: How exactly does the brain change when we learn to behave in a new way?

Parents marvel at the speed of their babies' development, excitedly celebrating every new action or word as a wondrous occasion. The infant brain grows rapidly, reaching close to adult size by two years of age. This suggests a simple theory: Perhaps learning is nothing more than brain growth, and children can be made smarter by enhancing this growth.

This theory goes back yet again to the phrenologists. Johann Spurzheim argued that mental exercise could enlarge cortical organs, much as muscles bulk up after physical training. Based on this theory, Spurzheim went on to develop an entire philosophy of education for both children and adults.

More than a century passed before his theory was finally tested scientifically. By that time, psychologists had invented a way of studying the effects of stimulation on the animal mind. Laboratory rats were placed in two different environments, one dull and the other "enriched." In the dull cage, a solitary rat lived with food and water containers as the only decoration. In the enriched cage, many rats lived together in a group and were provided daily with new toys. By running the rats through simple mazes, researchers found that the enriched rats were smarter. Presumably their brains were different, but exactly how?

In the 1960s Mark Rosenzweig and his colleagues decided to find out. Their method was startlingly simple: They weighed the cortex. It turned out that the enriched cage slightly enlarged the cortex on average. This was the first demonstration that experience causes the brain's structure to change.

You might not be surprised. After all, what about those MRI studies showing that London taxi drivers, musicians, and bilinguals have enlarged brain regions? Once again we must be careful not to read too much into statistical findings. The MRI studies showed correlation, but they did not prove causation.

Did driving a taxi, playing a musical instrument, and speaking a second language *cause* the brain to enlarge, as in Spurzheim's theory? Causation could be claimed if the brains of musicians and non-musicians were the same before musical training, and only became different afterward. But since the MRI study only collected data about "after," it cannot rule out an alternate interpretation: Perhaps some people are born with a brain enlargement that endows musical talent, and these gifted people are more likely to become musicians. Enlarged brains cause musical training, not the other way around.

Musicians may be selected on the basis of innate talent by music teachers and competitions. And musicians may be self-selecting, since people generally prefer activities at which they excel. This sort of problem, known as *selection bias*, complicates the interpretation of many statistical studies. Rosenzweig eliminated selection bias by *randomly* installing some rats in the enriched cage and others in the dull cage. This ensured that the two groups of rats started out statistically identical, enabling him to interpret any differences after the cage experience as having been caused by it.

For an even more direct demonstration of causation, one can use MRI to compare human brains before and after an experience. In this way, researchers found that learning to juggle balls thickened the cortex in the parietal and temporal lobes. And intensive study for exams caused the parietal cortex and the hippocampus to enlarge in medical students.

These results are impressive, but they are still not what we want. It is not enough to show that experience changes the brain. We also want to know whether the change is the cause of the improved performance. To understand why proof is still lacking, consider the following analogy. Imagine that musical training causes musicians to become more obese by forcing a sedentary lifestyle of practicing all day long. It would be wrong to conclude that obesity causes their improved musical performance. Similarly, showing that musical training enlarges the brains of musicians does not prove that this growth causes them to play their instruments better.

Rosenzweig showed that living in the enriched cage made rats smarter and also thickened their cortex. He did not prove, however, that it was the thickening that caused the improvement in intelligence. In fact, this seems unlikely, given what we know about the functions of cortical regions. The frontal lobe is thought to be important for skills like maze-running, but it showed little or no increase in size. The occipital lobe, which is responsible for visual perception, showed the largest increase.

In the end, we cannot equate cortical thickening with learning. We can say only that these two phenomena are correlated. Furthermore, the correlation is weak, once again revealed only by averages over groups. Cortical thickening is not a reliable predictor of learning in individuals.

Perhaps studying maze-running or juggling is the wrong approach. Maybe we should study more dramatic changes. Immediately after a stroke, for example, a patient usually experiences weakness or paralysis and may also lose speech and other mental abilities. Many patients improve dramatically over the next few months. What happens to the brain during recovery? Research on this question is of clear practical importance, as it could help us develop better therapies.

Strokes are caused by blocked or leaking blood vessels that damage the brain. The symptoms often indicate which side of the brain has been damaged. If patients struggle to control one side of the body, as is frequently the case, it means that the opposite side of the brain has been damaged, because each side of the brain controls muscles on the opposite side of the body. Neurologists can sometimes further pinpoint the affected brain region. To describe the location of cortical injury, a neurologist may specify a lobe or, if more precision is needed, a particular fold in a lobe. The folds have fancy-sounding names like "superior temporal gyrus," which means the uppermost fold in the temporal lobe. Alternatively, a cortical area may be specified by a number rather than a name, using a map published by the German neuroanatomist Korbinian Brodmann in 1909 (see Figure 11). In this book I will use the term *area* to mean a subdivision of Brodmann's map, and *region* to refer to any subdivision of the brain.

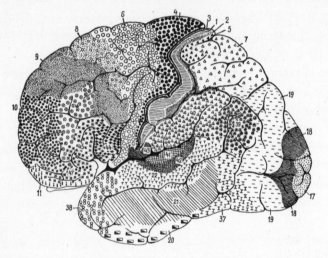

FIGURE 11. *Brodmann's map of the cortex*

Loss of movement after a stroke can result from damage to areas 4 and 6. Area 4 is the rearmost strip of the frontal lobe, just in front of the central sulcus, and area 6 is in front of area 4. Both are known to be important for control of movement. Language, too, is commonly impaired by stroke. That's a sign of damage to Broca's region (areas 44 and 45) or Wernicke's region (the back end of area 22), both in the left hemisphere.

Friends and family desperately want to know how much recovery is possible. Will Grandpa walk again? Will he talk? Movement tends to improve over time, but not much more after three months. Language also recovers most rapidly during the first three months, though it can continue to improve for months or years afterward. Neurologists know the three-month mark is important, but they do not know exactly why. More fundamentally, they don't know exactly what changes are taking place inside the brain as the patient recovers.

Obviously, the affected brain region might recover part or all of its function. But some cells near the malfunctioning blood vessel actually die, causing irreversible damage. Could the spared regions take over for the damaged region? Imagine that one of the players on a soccer team suffers an injury and is carried in agony off the playing field. There are no substitute players sitting on the bench, so the shorthanded (or shortfooted) team now plays worse. But as the game proceeds, the remaining players may adapt to the situation. If their comrade played in an attacking position before the injury, the defenders might compensate by starting to double as attackers.

So this is an important question: Can a cortical area acquire a new function after brain injury? There is some evidence for this after stroke, but stronger confirmation comes from cases of brain damage in early life. The disorder of epilepsy is defined by repeated spontaneous "seizures," or episodes of excessive neural activity. Children with very frequent and debilitating seizures are sometimes treated by removing one hemisphere of the cerebrum entirely. This is one of the most radical neurosurgical procedures, and it's astonishing that most children recover very well from it. Afterward they walk and even run, though movements of the hand on the opposite side are impaired. Their intellectual abilities are generally intact, and can even improve after surgery if the seizures are successfully eliminated.

One might argue that the recovery after hemispherectomy is not so surprising. Perhaps it's like losing a kidney. The remaining kidney need not do anything different; it just performs more of the same. But remember that some mental functions are lateralized, so the left and right sides of the brain are not equivalent. Because the left hemisphere specializes in language, its removal almost invariably leads to aphasia in adults. This is not true for children; linguistic functions migrate to the right hemisphere, demonstrating that cortical areas can indeed change their functions.

Given what we know about localization, it's not surprising that neurologists can guess the location of brain injury from the symptoms. Here's the surprising "yes, but": There may be a map dividing the cortex into areas with distinct functions, but the map is not fixed. The injured brain can redraw it.

The remapping of the cortex seen after stroke or surgery is more dramatic than the thickening reported by the neo-phrenologists. Can remapping also happen in healthy brains? Once again, insight can be gained from cases of severe injury—but to the body, not the brain. The following passage comes from an article by the neuroscientist Miguel Nicolelis:

> One morning in my fourth year of medical school, a vascular surgeon at the University Hospital in São Paulo, Brazil, invited me to visit the orthopedics inpatient ward. "Today we will talk to a ghost," the doctor said. "Do not get frightened. Try to stay calm. The patient has not accepted what has happened yet, and he is very shaken."
>
> A boy around 12 years old with hazy blue eyes and blond curly hair sat before me. Drops of sweat soaked his face, contorted in an expression of horror. The child's body, which I now watched closely, writhed from pain of uncertain origin. "It really hurts, doctor; it burns. It seems as if something is crushing my leg," he said. I felt a lump in my throat, slowly strangling me. "Where does it hurt?" I asked. He replied: "In my left foot, my calf, the whole leg, everywhere below my knee!"
>
> As I lifted the sheets that covered the boy, I was stunned to find that his left leg was half-missing; it had been amputated right below the knee after being run over by a car. I suddenly realized that the child's pain came from a part of his body that no longer existed. Outside the ward I heard the surgeon saying, "It was not him speaking; it was his phantom limb."

Modern methods of amputation were invented in the sixteenth century by Ambroise Paré, who perfected his art as a surgeon for the French army. Paré was born at a time when surgery was performed by barbers, because it seemed like a crude act of butchery too lowly for physicians. Working on the battlefield, Paré learned how to tie off large arteries to prevent amputees from bleeding to death. He eventually earned employment with several French kings and a place in the history books as the "father of modern surgery."

Paré was the first to report that amputees complained of an imaginary limb still attached to the body where the real limb used to be. Centuries later, the American physician Silas Weir Mitchell coined the term *phantom limb* to describe the same phenomenon in Civil War soldiers. His many case studies established that phantom limbs are the rule, not the exception. Why had they gone unremarked for so long? Before the surgical innovations of Paré, very few people survived amputation, and the complaints of those who did may have been dismissed as mere delusions. But far from being irrational, amputees are well aware that the phantom is not real, and because its sensations are usually painful, they beg doctors to make it go away.

Along with naming it, Mitchell proposed a theory to explain the phenomenon. He suggested that irritated nerve endings in the stump were sending signals to the brain, which interpreted them as sensations from the missing limb. Inspired by the theory, some surgeons tried amputating the stump, but this didn't help. Today many neuroscientists believe a different theory: Phantom limbs are caused by a remapping of the cortex.

The reorganization is not of the entire cortex; it's thought to be confined to a particular area. We previously learned about area 4, the strip in front of the central sulcus that controls movement. Just behind the central sulcus is area 3, which is involved in the bodily sensations of touch, temperature, and pain. In the 1930s the Canadian neurosurgeon Wilder Penfield mapped both areas in his patients by using electrical stimulation. After opening the skull to expose the brain for epilepsy surgery, Penfield applied his electrode to different locations in area 4. Each stimulation caused some part of the patient's body to move. Penfield drew the correspondence between area 4 locations and body parts (Figure 12, right), calling the map a "motor homunculus." (*Homunculus* is from the Latin for "little human.") Likewise, after each stimulation of area 3, the patient reported feeling a sensation in some part of the body. Penfield mapped the "sensory homunculus" in area 3 (left), and it looked similar to the motor one. Both ran in parallel along opposite banks of the central sulcus. (Roughly speaking, these maps represent vertical planes passing through the brain from ear to ear. The plane of the sensory map is just behind the central sulcus, and that of the motor map just in front. Only the outer border is cortex; the rest is the interior of the cerebrum.)

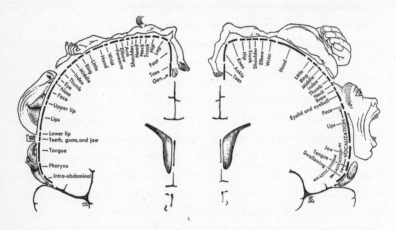

FIGURE 12. *Functional maps of cortical areas 3 and 4: the "sensory homunculus"* (left) *and the "motor homunculus"* (right)

The face and hands dominate the maps, even though they are small parts of the body. Their cortical magnification reflects their disproportionate importance in sensation and movement. Could the sizes of their territories be changed by amputation, which suddenly reduces the importance of a body part to zero? Using such reasoning, the neurologist V. S. Ramachandran and his collaborators have proposed that phantom limbs are caused by remapping of area 3. If the lower arm is amputated, its territory in the sensory homunculus loses its function. The surrounding territories, dedicated to the face and upper arm, encroach upon the nonfunctional one by advancing their borders. (You can see the adjacencies in Penfield's drawing.) These two intruders start to represent the lower arm as well as their original body parts, giving the amputee the sensation of a phantom limb.

According to the theory, the remapped face territory should represent the lower arm as well as the face. Therefore Ramachandran predicted that stimulation of the face would cause sensations in the phantom limb. Indeed, when he stroked the face of an amputee with a Q-tip, the patient reported feeling sensations not only in his face but also in his phantom hand. The theory likewise predicts that the remapped upper-arm territory should represent the lower arm as well as the upper arm. When Ramachandran touched the stump, the patient felt sensations in both the stump and his phantom hand. These ingenious experiments strikingly confirmed the theory that amputation caused remapping of area 3.

Ramachandran and his collaborators used technology no more advanced than a Q-tip. In the 1990s an exciting new method of brain imaging was introduced. Functional MRI revealed every region's "activity," or how much that part of the brain was being used. By now the images of functional MRI (fMRI) are familiar from their frequent appearance in the news media. They are usually shown superimposed on regular MRI images. The black-and-white MRI image shows the brain, and laid on top are the colored blotches of the fMRI image, which indicate the active regions. You can always recognize fMRI+MRI as "spots on brains," while MRI is just brains.

Researchers imaged volunteers while they performed mental tasks in the laboratory. If a task activated a region, causing it to "light up" in the image, that was a clue to the region's function. Neurology had always been hampered by the accidental nature of brain lesions, but fMRI enabled precise and repeatable experiments on localization of function. Brodmann's map became indispensable as researchers worked hard to assign functions to each of its areas. The boom in scientific papers spurred many universities to invest large sums of money in fMRI machines, or "brain scanners."

Researchers also repeated Penfield's mapping of the sensory and motor homunculi. They observed which locations in area 3 were activated by touching parts of the body, and which locations in area 4 were activated when the subject moved parts of the body. It was thrilling to reproduce Penfield's maps with fMRI rather than his crude method of opening up the skull. Researchers also studied remapping, verifying Ramachandran's claim of a downward shift of the face representation in area 3 of amputees. As the theory predicted, the shift occurred only in those amputees who experienced phantom limb pain, not in pain-free amputees.

Amputation may not be injury to the brain, but it's still a highly abnormal kind of experience. Do brains remap in more normal forms of learning? Violinists and other string musicians use the left hand to finger the strings of their instruments. Studies show enlargement of the left-hand representation within area 3, which is likely due to extensive musical practice. It's impressive that fMRI can not only assign functions to Brodmann areas but also resolve fine changes within a single area. This research is far more sophisticated than studies of total brain size like Galton's. It is bound to tell us more interesting things about cortical remapping, and it may even be useful for understanding crippling disorders of movement that seem to be caused by too much practice. Such disorders, known as focal dystonias, have tragically ended the careers of brilliant musicians.

Explaining learning in terms of the expansion of cortical areas or subareas, however, is still in the spirit of phrenology. It's not so different in concept from the studies of cortical thickening, and the correlations are still statistically weak. The approach may be powerful, but it has limitations. For example, studies of Braille readers also show an enlarged hand representation. The remapping approach cannot easily distinguish between learning violin and Braille, which are two very different skills. And even if this particular problem can be solved, the general difficulty will remain.

Researchers have one other way of studying changes in the brain, which does not depend on the concept of remapping. Using fMRI, they have attempted to find differences in the level of activation of brain regions. For example, they have reported lower activation of the frontal lobe in schizophrenics performing certain mental tasks. At the moment such correlations are statistically weak, but this intriguing line of research may well tell us much about brain disorders and possibly lead to superior methods of diagnosing them.

At the same time, fMRI studies may have a fundamental limitation. Brain activation changes from moment to moment, roughly as quickly as thoughts and actions change. To find the cause of schizophrenia, we must identify some brain anomaly that is constant. Suppose that your car starts to shake whenever you drive faster than 30 miles per hour and turn the steering wheel to the right. This behavior is intermittent, so it's only a symptom. It's caused by something wrong with your car at a more basic level. Noticing symptoms is crucial, but it's only the first step toward identifying the underlying cause.

Why are we still trying to use phrenology to explain mental differences? It's not because the strategy is good. It's because we have failed to come up with a better one. Do you know the joke about the policeman who comes upon a drunk crawling on the ground near a lamppost? The drunk explains, "I lost my keys around the corner." The policeman asks, "Well, why don't you search over there?" The drunk replies, "I would, but there's more light under the lamppost." Like the drunk who works with what he's got, we know that size reveals little about function, but we look at it anyway because that's what we can see with existing technologies.

To understand the failings of phrenology, can we compare with a more successful example of relating function to size? Instead of investigating whether brainy people are smarter, let's ask whether brawny people are stronger. The size of a muscle can be measured via MRI, and its strength with a machine that looks like one in the weight room at your health club. Researchers have found correlation coefficients ranging from 0.7 to 0.9, which is much stronger than the correlation between brain size and IQ. Muscle size accurately predicts strength, just as we'd expect.

Why are size and function so closely related for muscles but not for brains? Think of a muscle as operating like a factory in which all workers do the same thing. If every worker singlehandedly performs all the steps required for making an entire widget, doubling the size of the workforce will double the factory's output of widgets. Likewise, every fiber of a muscle performs the same task. All the fibers are lined up in parallel, and all pull in the same direction. Their contributions to the force are additive (you can simply add them together to get the total), so a muscle with more fibers should be stronger.

Now consider a factory with a more complex organization. Each worker performs a different task, like fastening a screw or welding a joint. To make even a single widget, all the workers must cooperate. Economists say that such division of labor is efficient because specialization allows each worker to become highly skilled at each task. However, doubling the number of workers will likely fail to double the output of widgets. It's not easy to integrate the new workers into the existing organization in a way that increases output. In fact, adding more workers could even reduce output by disrupting the workflow. As Brooks' Law — a maxim of software engineers — puts it, "Adding more programmers to a late software project makes it later."

The brain works like the more complex factory. Each of its neurons performs a tiny task, and they cooperate in intricate ways to carry out mental functions. That's why performance depends less on the number of neurons and more on how they are organized.

The factory analogy explains the limitations of phrenology. Can it also explain remapping? The American neuropsychologist Karl Lashley believed that mental functions were widely distributed across the cortex, and charged that most of the boundaries of Brodmann's map were figments of the imagination. Nevertheless, this archenemy of localizationism could not completely deny the experimental evidence in its favor. In 1929 he countered with his doctrine of cortical *equipotentiality*. Lashley granted that every cortical area is dedicated to a specific function, but every area also has the *potential* to assume some other function, he claimed.

Returning to our imaginary factory — the more complex one — let's suppose that a worker is reassigned to a new task. The initial clumsiness will eventually give way to proficiency. Workers may be specialized, but they are also equipotential. When provided with new inputs, they can change their functions.

Lashley's doctrine has some element of truth but is too sweeping. The cortex is not infinitely adaptable. If it were, every stroke patient would recover completely. To understand the limits of adaptation and develop ways to enhance it, we need a deeper understanding. We know that the cortex can remap, but how exactly does the function of an area change?

We can't answer this without addressing a more basic issue: What defines the function of a cortical area in the first place? Broca's and Wernicke's regions are dedicated to language, and Brodmann areas 3 and 4 are dedicated to bodily sensation and movement. But *why* these functions? And how are they executed?

It's hopeless to answer these questions by studying only brain regions, their sizes, and their activity levels. We must look at the organization of the brain on a much finer scale. A cortical area can contain over 100 million neurons. How are they organized to perform mental functions? In the next few chapters we'll explore this question, along with the idea that brain function depends heavily on the *connections* between neurons.

PART II

CONNECTIONISM

THREE

No Neuron Is an Island

THE NEURON IS my second-favorite cell. It's a close runner-up to my favorite: sperm. If you have never looked into a microscope to see sperm swimming furiously, grab your favorite biologist by the lapels of his or her lab coat and demand a viewing session. Gasp at the urgency of their mission. Mourn their imminent death. Marvel at life stripped down to its bare essentials. Like a traveler with a single small suitcase, a sperm carries little. There are mitochondria, the microscopic power plants that drive the whipping motion of its tail. And there is DNA, the molecule that carries the blueprint of life. No hair, no eyes, no heart, no brain — nothing extraneous comes along for the ride. Just the information, please, written in DNA with the four-letter alphabet A, C, G, and T.

If your biologist friend is still game, ask to see a neuron. Sperm impress by their unceasing motion, but a neuron takes your breath away with its beautiful shape. Like a typical cell, a neuron has a boring round part, which contains its nucleus and DNA. But this *cell body* is only a small part of the picture. From it extend long, narrow branches

FIGURE 13. *My favorite cells: sperm fertilizing an egg* (left) *and a neuron* (right)

that fork over and over, much like a tree. Sperm are sleek and minimalist, but neurons are baroque and ornate (see Figure 13).

Even in a crowd of 100 million, a sperm swims alone. At most one will achieve its mission of fertilizing the egg. The competition is winner take all. When one sperm succeeds, the egg changes its surface, creating a barrier that prevents other sperm from entering. Whether brought together by a happy marriage or a sordid affair, sperm and egg form a monogamous couple.

No neuron is an island. Neurons are polyamorous. Each embraces thousands of others as their branches entangle like spaghetti. Neurons form a tightly interconnected network.

The sperm and the neuron symbolize two great mysteries: life and intelligence. Biologists would like to know how the sperm's precious cargo of DNA encodes half the information required for a human being. Neuroscientists would like to know how a vast network of neurons can think, feel, remember, and perceive — in short, how the brain generates the remarkable phenomena of the mind.

The body may be extraordinary, yet the brain reigns supreme in its mystery. The heart's pumping of blood and the lung's intake of air remind us of the plumbing in our houses. They may be complex, but they do not seem mysterious. Thoughts and emotions are different. Can we really understand them as the workings of the brain?

A journey of a thousand miles begins with a single step. To understand the brain, why not start with its cells? While a neuron may be a kind of cell, it is far more complex than any other. This is most obvious from its profuse branches. Even after many years of studying neurons, I am still thrilled by their majestic forms. I'm reminded of the mightiest tree on earth, the California redwood. Hiking in Muir Woods, or other redwood forests on the Pacific coast of North America, is a good way to feel small. You see trees that live for centuries or even millennia, enough time to grow to vertiginous heights.

Am I overreaching to compare a neuron to the towering redwood? In absolute size, yes, but consider further how these wonders of nature stack up against each other. The redwood's twigs are as thin as one millimeter, a width 100,000 times smaller than the tree's football-field height. A branch of a neuron, called a *neurite,* can extend from one side of the brain to the other, yet can also narrow to 0.1 micrometer in diameter. These dimensions differ by a factor of one *million.* In its relative proportions, a neuron puts a redwood to shame.

But why do neurons have neurites? And why do they branch to look like trees? In the case of a redwood, the reason for branches is obvious: The redwood's crown captures light, which is a source of energy. A passing sunbeam will almost surely collide with a leaf rather than travel all the way to the ground. Likewise, a neuron is shaped to capture contacts. If a neurite passes through the branches of another neuron, it will likely collide with one of them. Just as a redwood "wants" to be struck by light, a neuron "wants" to be touched by other neurons.

Every time we shake hands, caress a baby, or make love, we may be reminded that human life depends on physical contact. But why do neurons touch? Suppose that the sight of a snake causes you to turn and run. You respond because your eyes are able to communicate a message to your legs: *Move!* That message is conveyed by neurons, but how?

Neurites are much more densely packed than the branches of a forest or even a tropical jungle. Think instead of a plate of spaghetti — or microscopically fine capellini. Neurites entangle much like the jumbled strands on your plate, allowing one neuron to touch many others. Where two neurons touch, there can be a structure called a *synapse,* a junction through which the neurons communicate.

But contact alone does not make a synapse, which most commonly transmits chemical messages. A molecule known as a *neurotransmitter* is secreted by the sending neuron and sensed by the receiving neuron. Secretion and sensing are performed by still other types of molecules. The presence of such molecular "machinery" signifies that a contact point is actually a synapse, as opposed to a place where one neurite just goes past another.

These telltale signs are blurred in an ordinary microscope, which uses light to make images, but show up nicely with a more advanced microscope based on electrons rather than light. The image shown in Figure 14 is a highly magnified (100,000×) view of a cut through brain tissue. There are two large, round cross-sections of neurites (marked "ax" and "sp"). These are like the cut ends of strands that would be exposed if you sliced through spaghetti. The arrow points to a synapse between the neurites, which are separated by a narrow cleft. Now we see that the term *contact point* is not entirely accurate, as the neurites come extremely close to each other but do not really touch.

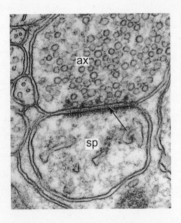

FIGURE 14. *A synapse in the cerebellum*

On either side of the cleft is the molecular machinery for sending and receiving messages. One side is dotted with many little circles, tiny bags called vesicles that store neurotransmitter molecules ready for use. On the other side the membrane holds a dark fuzz called the *post-synaptic density*, which contains molecules known as *receptors*.

How does this machinery transmit a chemical message? The sender secretes by dumping the contents of one or more vesicles into the cleft. The neurotransmitter molecules spread out in the salty water there. They are sensed by the receiver when they encounter receptor molecules embedded in the postsynaptic density.

Many types of molecule are used as neurotransmitters. Each is assembled from atoms bonded to each other, as in the examples shown in Figure 15. (In these "ball-and-stick" models, each ball represents an atom and each stick a chemical bond.) You can see that each type of neurotransmitter has a characteristic shape determined by the specific arrangement of its atoms, a fact that will become important shortly.

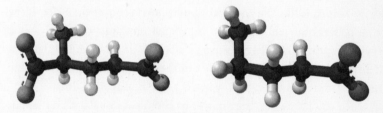

FIGURE 15. *"Ball-and-stick" models of neurotransmitters: glutamate (left) and GABA (right)*

On the left is the most common one, glutamate. This is best known to the public in the form of monosodium glutamate (MSG), which is used as a flavor enhancer in Chinese and other Asian cuisines. Few realize that glutamate also plays a crucial role in brain function. Shown on the right is the second most common, gamma-aminobutyric acid, or GABA for short.

More than one hundred neurotransmitters have been discovered so far. The list sounds long. Do you ever feel overwhelmed in the liquor store, when you see the shelves stocked with so many brands of beer and wine? If you're a creature of habit, you might buy the same one or two brands every time and serve them to your friends at every party you give. That's what neurons do. With few exceptions, a neuron secretes the same small set of neurotransmitters — often only a single neurotransmitter — at all of its synapses. (The synapses in question are those made by a neuron onto others, not those received by a neuron.)

Now let's consider receptor molecules, which are much larger and more complex than neurotransmitters. Part of each molecule sticks out from the surface of the neuron, like the head and arms of a kid using an inner tube to float on water. This protrusion is the part of the receptor that senses neurotransmitter.

A glutamate receptor senses glutamate but ignores GABA and other neurotransmitters. Likewise, a GABA receptor senses GABA but ignores other molecules. Where does this specificity come from? Think of a receptor as a lock and the neurotransmitter as a key. As we saw above, each type of neurotransmitter has a distinctive molecular shape, which is like the pattern of bumps and grooves on a key. Every type of receptor has a location called the binding site, which has a characteristic shape like the innards of a hole in a lock. If the shape of the neurotransmitter matches that of the binding site, it activates the receptor, much as the right key in the right lock opens a door.

Once you know that the brain uses chemical signals, it's no longer surprising that drugs can alter the mind. A drug is a molecule too, and can be shaped like a neurotransmitter. If the mimicry is faithful enough, the drug will activate receptors, much as a copy of a key can open the same lock as the original. Nicotine, the addictive chemical in cigarettes, activates receptors for the neurotransmitter called acetylcholine. Other drugs inactivate receptors, much as an inaccurate copy of a key might turn partially and jam the lock. Phencyclidine or PCP, known on the street as "angel dust" in honor of its recreational use for hallucinogenic effects, inactivates glutamate receptors.

It's worth pausing to consider how we usually perceive secretions. Spit. Sweat. Urine. We suppress the urge to expectorate in polite company, plug glands with antiperspirants, and flush toilets in quiet privacy. We are embarrassed by secretions, reminders of our flesh and blood. Surely they live in a world apart from entities as ethereal and refined as our thoughts. But the truth is more shocking: The mind depends on an untold number of microscopic emissions. The brain secretes thoughts!

It may seem strange that neurons communicate with chemicals, but we humans do it too. Granted, we rely much more on language or facial expressions. But occasionally we signal each other with smells. While the message of aftershave or perfume is open to interpretation, something along the lines of "I'm sexy" or "Come hither" is a safe guess. Other animals don't have to purchase smell in a bottle. A female dog in heat naturally secretes a chemical signal called a pheromone, which wafts through the neighborhood to bring droves of male dogs by their noses.

Such chemical messages express desire more primitively than Shakespeare's love sonnets. Then again, so do poems that start with "Roses are red, violets are blue." We should distinguish between the medium and the message. Is there something fundamentally primitive about chemical signals as a medium for communication? There are indeed several limitations, but the brain has found a way to circumvent all of them.

Chemical signals are typically slow. If a woman walks into a room, you will usually hear her footsteps and see her clothing well before you catch a whiff of her perfume. A draft in the room might blow the scent toward you more rapidly, but it will still arrive more slowly than sound and light. Nervous systems, however, generate speedy reactions. When you suddenly jump away from a car piloted by a reckless driver, your neurons signal each other quickly. How can they accomplish this with chemical messages? Think of it this way: Even the slowest runner can finish a race in the blink of an eye if the racetrack is just a few strides long. Though chemical signals may move slowly, the distance that they have to travel across the synaptic cleft is extremely short.

Chemical signals also seem crude because it is difficult to send them to specific targets. All the partygoers surrounding a woman can smell her perfume. Wouldn't it be more romantic if her fragrance could be sensed only by her beloved? Alas, no inventor has managed to create a scent that is focused in this way. So what keeps the chemical messages at one synapse from spreading like perfume and being sensed by others? The answer is that a synapse "recycles" neurotransmitter by sucking it back up, or degrading it into an inert form, leaving the molecules with little chance to wander. It's no trivial matter for the nervous system to minimize crosstalk — as engineers call the spreading phenomenon — because synapses are packed so close to each other. With a billion synapses to a cubic millimeter, the brain is far more crowded than Manhattan, and that island's residents often complain about hearing conversations (and much else) from each other's apartments.

Finally, the timing of chemical signals is not easily controlled. A woman's perfume may linger in a room long after she has left the party. The dawdling of neurotransmitter is averted by the same mechanisms of recycling and degradation that squelch crosstalk. This allows chemical messages between neurons to occur at precise times.

These properties of synaptic communication — speed, specificity, and temporal precision — are not shared by other types of chemical communication inside your body. After you jump away from the car in the street, your heart races, you breathe heavily, and your blood pressure skyrockets. This is because your adrenal gland secreted adrenaline into your bloodstream, which was sensed by cells in your heart, lungs, and blood vessels. The reactions of the "adrenaline rush" may seem immediate, but actually they are tardy. They happened *after* you jumped away from the car, because adrenaline spreads through your bloodstream more slowly than signals jump from neuron to neuron.

Secretion of hormones into the blood is the most indiscriminate type of communication, called broadcasting. Just as a television show is received by many households, and a perfume by everyone in a room, a hormone is sensed by many cells in many organs. In contrast, communication at a synapse is restricted to the two neurons involved, just as a telephone call connects the two people on the line. Such point-to-point communication is much more specific than broadcasting.

In addition to chemical signals between neurons, there are also electrical signals in the brain. These travel *within* neurons. Neurites contain salty water rather than metal, but they nonetheless resemble, in both form and function, the telecom wires that crisscross the planet. Electrical signals can travel long distances by propagating through neurites, much as they move along wires. (Interestingly, the mathematical equations developed by Lord Kelvin in the nineteenth century to describe electrical signals in undersea telegraph cables have been used in the modeling of neurites.)

In 1976 the legendary engineer Seymour Cray unveiled one of the most famous supercomputers in history, the Cray-1 (see Figure 16). Some called it the "world's most expensive loveseat," and indeed its sleek exterior could have graced the living room of a 1970s playboy. Its interior was anything but sleek, containing 67 miles of tangled wire in lengths spanning 1 to 4 feet. This looked like a chaotic mess to the casual observer, but actually it was highly ordered. Every wire transmitted information between a specific pair of points chosen by Cray and his design team from locations on thousands of "circuit boards" holding silicon chips. As is common in electronic devices, the wires were wrapped with insulating material to prevent crosstalk.

You may think the Cray-1 looks complex, but it's laughably simple compared with your brain. Consider that *millions* of miles of gossamer neurites are packed inside your skull, and they are branched rather than straight like wires. The tangle in your brain is far worse than that of the Cray-1. Nevertheless, the electrical signals in different

FIGURE 16. *The Cray-1 supercomputer, exterior* (left) *and interior* (right)

neurites — even adjacent ones — interfere with each other very little, just as in insulated wires. Transmission of signals between neurites occurs only at specific points, those junctions called synapses. Similarly, signals cross from one wire to another in the Cray-1 only at locations where the insulation is removed and the metals come directly into contact.

I've spoken of neurites generically up to now, but many neurons have two types of neurite — dendrites and axon. The dendrites are shorter and thicker. Several emanate from the cell body and branch in its vicinity. A single axon, long and thin, travels far from the cell body and branches out at its destination.

Dendrites and axons not only look different but play different roles in chemical signaling. Dendrites are on the receiving end of synapses. Their membranes contain the receptor molecules. Axons send signals to other neurons by secreting neurotransmitter at synapses. In other words, the typical synapse is from axon to dendrite.

The electrical signals of dendrites and axons also differ. In axons, electrical signals are brief pulses known as *action potentials,* each lasting about a millisecond (see Figure 17). Action potentials are informally known as "spikes," owing to their pointy appearance, so let's use this nickname for convenience. Neuroscientists often say, "The neuron spiked," much as a financial reporter writes, "The stock market spiked on bank profits." When a neuron spikes, it is said to be "active."

Spikes are reminiscent of Morse code, which you've probably heard in old movies as a sequence of long and short pulses generated by a

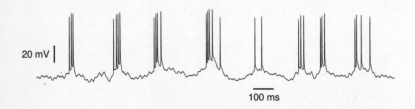

20 mV |

100 ms

FIGURE 17. *Action potentials, or "spikes"*

telegraph operator pressing a lever. In early telecom systems, pulses were just about the only type of signal that could be heard clearly above the static. Signals tend to become more corrupted by noise as they travel farther. That's why Morse code was still used for long-distance communication even decades after the telephone became popular for local calls. Nature "invented" the action potential for much the same reason, to transmit information over long distances in the brain. Thus spikes occur mainly in the axon, the longest type of neurite. In small nervous systems like that of *C. elegans* or a fly, neurites are shorter and many neurons do not spike.

• So how are these two types of neural communication, chemical and electrical, related? Simply put, a synapse is activated when a passing spike triggers secretion. On the other side of the synapse, receptors sense neurotransmitter and then make electrical current flow. In more abstract terms, a synapse converts an electrical signal into a chemical signal and then back into an electrical signal.

Conversion between signal types is common in our everyday technologies. Imagine two people conversing by telephone. Electrical signals travel between them along a continuous wire. (Let's ignore the fact that modern telephone networks additionally use light signals in optical fibers.) But electrical signals do not traverse the narrow gap of air between the handset and the ear; instead, they are converted into acoustic signals. After a journey of a thousand miles as electricity, it is sound that makes the leap to the listener's eardrum. Similarly, an electrical signal may travel far in the brain along an axon, but it does not reach the next neuron directly. Rather, it is converted into a chemical signal, which jumps across the synaptic cleft to the other neuron.

If one neuron can signal a second neuron through a synapse, the second neuron can signal a third, and so on. A sequence of such neurons is known as a *pathway*. This is how neurons can communicate with one another even if they are not directly connected by a synapse.

Unlike the mountain paths that we hike, neural pathways are directional. This is because synapses are one-way devices. When there is a synapse between two neurons, we say that they are connected to each other, like two friends talking on the telephone. But the metaphor is flawed, because a telephone transmits information in both directions. At any given synapse, the messages travel one way: One neuron is always the sender, the other always the receiver. This is not because one neuron is "talkative" or the other "taciturn." Rather, it has to do with the structure of the synapse. The machinery for secreting neurotransmitter is on one side and that for sensing neurotransmitter on the other.

In principle, neurites are two-way devices along which electrical signals can travel in either direction. In practice, a spike normally travels along an axon away from the cell body, and electrical signals travel along dendrites toward the cell body. Synapses impose this directionality onto neurites. In your circulatory system, blood flows in your veins toward your heart. If a vein were simply a tube, blood could potentially flow in either direction. But a vein also contains valves, which prevent blood from flowing backward. Valves impose directionality on veins in much the same way that synapses impose it on neural pathways.

So a pathway in the nervous system is defined by stepping across synapses from neuron to neuron, respecting the direction of each synapse (see Figure 18). Inside one neuron, electrical signals flow from dendrites to cell body to axon. Chemical signals jump from the axon of this neuron to the dendrite of another neuron. Inside this neuron, electrical signals again flow from dendrites to cell body to axon. They are converted into chemical signals to jump to another neuron, and the process continues. Because the synaptic cleft is extremely narrow, almost all of the distance spanned by the pathway is actually within neurons rather than between neurons. Furthermore, most of this distance runs through axons, which are much longer than dendrites.

If you've eaten poultry, you may have spied bundles of axons on your dinner plate. They are called nerves, and can be recognized as soft whitish strings. They are not to be confused with tendons, which are tougher, or blood vessels, which are darker. Dissecting an uncooked nerve with a very sharp tool causes it to fray, much as a rope unravels into many threads when cut. The "threads" of a nerve are its axons.

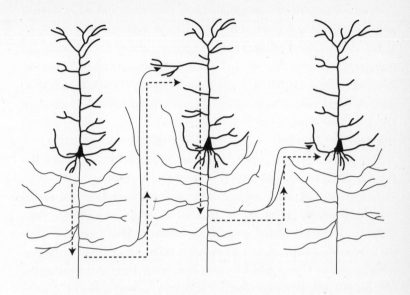

FIGURE 18. *Multineuron pathway in the nervous system*

Nerves are rooted to the surface of the brain or spinal cord, together known as the central nervous system (CNS). Because most nerves extend and branch toward the surface of the body, they are known as the peripheral nervous system (PNS). The axons in nerves come from cell bodies in the CNS or in little outposts of neurons known as peripheral ganglia. The CNS and the PNS together make up the nervous system, defined alternatively as the set of all neurons and the cells that support them. The emphasis on nerves in the term *nervous system* is perhaps misleading, as the brain and spinal cord are its predominant parts.

Now let's return to the question posed earlier: How does the sight of a snake cause you to turn and run? The rough answer is that your eyes signal your brain, which signals your spinal cord, which signals your legs. The first step is mediated by the optic nerve, a bundle of a million axons from the eye to the brain. The second step happens through the pyramidal tract, a bundle of axons from the brain to the spinal cord. (A bundle of axons in the CNS is known as a tract rather than a nerve.) The third step passes through the sciatic and other nerves, which connect your spinal cord to your leg muscles.

Let's consider the neurons at the beginning and end of the pathways mediated by these axons. At the back of your eye is a thin sheet of neural tissue called the retina. Light from the snake strikes special neurons in the retina called photoreceptors, which respond by secreting chemical messages, which in turn are sensed by other neurons. More generally, every one of your sense organs contains neurons that are activated by some type of physical stimulus. Sensory neurons kick off the journey along neural pathways from stimulus to response.

These pathways end when axons in nerves make synapses onto muscle fibers, which respond to secretion of neurotransmitter by contracting. The coordinated contraction of many fibers causes a muscle to shorten and produce a movement. More generally, every one of your muscles is controlled by axons that come from motor neurons. The English scientist Charles Sherrington, who won a Nobel Prize in 1932 and coined the term *synapse*, emphasized that muscles are the final destination of all neural pathways: "To move things is all that mankind can do . . . for such the sole executant is muscle, whether in whispering a syllable or felling a forest."

Between sensory and motor neurons there are many pathways, some of which we will consider in detail in later chapters. It's clear that these pathways exist; if they didn't, you wouldn't be able to respond to stimuli. But exactly how do signals travel along pathways?

When California joined the United States in 1850, communicating with the eastern states took weeks. The Pony Express was created in 1860 to speed up mail delivery. Along its two-thousand-mile route from California to Missouri were 190 stations. A mailbag traveled day and night, switching horses at every station and changing riders every six or seven stations. After reaching Missouri, messages traveled by telegraph to states farther east. The total transit time for a message between the Pacific and the Atlantic was reduced from twenty-three to ten days. The Pony Express operated for only sixteen months before being completely replaced by the first transcontinental telegraph, which in turn was succeeded by telephone and computer networks. The technology may have changed, but the underlying principle has not: A communication network must have a means of relaying messages from station to station along pathways.

It's tempting to think of the nervous system as a communication network that relays spikes from neuron to neuron. A neural pathway would behave like dominoes, with each spike igniting the next spike in the pathway in the same way that each falling domino tips over the next one in the chain. This would explain how your eye tells your legs to move when you see a snake. But in fact it's not that simple. While it's true that an axon relays spikes from the cell body to synapses, it turns out that a synapse does not simply relay spikes to the next neuron.

Almost all synapses are weak. The secretion of neurotransmitter causes a tiny electrical effect in the next neuron, far below the level required to cause a spike. Imagine a chain of dominoes spaced too far apart. The falling of one won't have any effect on the next. Likewise, a single neural pathway cannot typically relay a spike on its own — but as I'll explain below, this is a good thing.

"Two roads diverged in a yellow wood / And sorry I could not travel both / And be one traveler, long I stood," wrote Robert Frost in "The Road Not Taken." A spike does not share Frost's dilemma when it comes to a fork in an axon. Not limited to being "one traveler," the spike duplicates itself, giving rise to two spikes that take both branches. By doing this repeatedly, a single spike starting near the cell body becomes many spikes that reach every branch of the axon, amplitude undiminished. All of the synapses made by the axon onto other neurons are stimulated to secrete neurotransmitter.

Through these outgoing synapses, neural pathways diverge like the roads in the poem. That's why stimulating one sense organ can cause multiple responses. The sight of a snake makes you want to run, because of pathways from your eyes to your legs. But the sight of a tasty steak causes your mouth to water, this time thanks to pathways from your eyes to your salivary glands. Because these two types of pathways diverge from the eyes, it's no mystery that either running or salivation is possible after you see something. The mystery is quite the opposite: Why is there only one response? If signals took all possible pathways, any stimulus would cause every muscle and gland to become activated, and clearly that doesn't happen.

The reason is that signals don't get through pathways so easily. We already saw that single synapses and pathways do not relay spikes. So how do signals ever get through? Although the branches of dendrites look similar to those of axons, their function is completely different. Axons diverge, but dendrites *converge.* Where two branches join, electrical currents can meet as they flow toward the cell body, and can combine like the water of merging streams. And as a lake collects water from many streams, the cell body collects currents from the many synapses converging onto its dendrites.

Why is convergence important? Although a single synapse is typically too weak to drive a neuron to spike, *multiple* converging synapses can do the job. If they are activated simultaneously, they can collectively "convince" a neuron to spike. Because a spike is "all or none," we can regard it as the output of a "neural decision." By this metaphor, I do not mean that a neuron is conscious or thinks in the same way that a human does. I simply mean that a neuron is not wishy-washy. There is no such thing as half a spike.

When we're deciding, we may seek advice from friends and family. Similarly, a neuron "listens" to other neurons through its converging synapses. The cell body sums the electrical currents, effectively tallying the votes of the "advisors." If the tally exceeds a threshold, the axon spikes. The value of this threshold determines whether a neuron decides easily or reluctantly, much as political systems can require a simple majority, a two-thirds majority, or unanimity.

In many neurons, the electrical signals of dendrites are continuously graded, unlike the all-or-none spikes of the axon. This is well suited for representing the entire range of possible vote tallies. A spike in the dendrites would be premature — like calling an election before all the votes are in. Only after the cell body tallies all the votes can spikes occur in the axon. If dendrites lack spikes, they cannot transmit information over long distances; that's the reason dendrites are much shorter than axons.

One of the basic slogans of a democracy is "One person, one vote." All votes are weighted equally, as in the neural model above. But we may be less democratic when combining the advice of our friends and family, giving more weight to some opinions than to others. Similarly, a neuron actually weights its "advisors" unequally. Electrical currents have magnitudes. Strong synapses produce large currents in the dendrite, and weak synapses produce small currents. The "strength" of a synapse quantifies the weight of its vote in the decision of a neuron. And it's possible for a neuron to receive multiple synapses from another neuron, as if allowing it to cast multiple votes — a further kind of favoritism.

We've arrived at the "weighted voting model" of a neuron. In any type of voting there is some requirement for simultaneity. In politics, this is achieved by asking everyone to go to the polls on a predetermined day. Since synapses can vote at any time, it's always election day in the brain. (Actually, the metaphor is slightly misleading — synaptic votes are tallied over a time period much shorter than a day, ranging from milliseconds to seconds.) The votes of two synapses are counted in the same tally only if their electrical currents are close enough in time to overlap.

Think of synaptic currents as insults being thrown at someone. Any single insult is too weak to excite a temper tantrum (a spike), so if the insults come only infrequently, the person won't get angry. But if there are many simultaneous insults or if they come in quick succession, they can add up — until the "last straw" pushes the person over the threshold.

In the explanation of neural voting I left out an important feature of synapses for the sake of simplicity. It turns out that "yes" votes are not the only kind tallied by neurons. Another kind of synapse registers "no" votes. The yes–no distinction arises because activation of a synapse causes current to flow, and two directions of flow are possible. *Excitatory* synapses say "yes" because they make electrical current flow *into* the receiving neuron, which tends to "excite" spiking. *Inhibitory* synapses say "no" because they make current flow *out* of the neuron, which tends to "inhibit" spiking.

Inhibition is crucial to the operation of the nervous system. Intelligent behavior is not just a matter of making appropriate responses to stimuli. Sometimes it's even more important to *not* do something—not reach for that doughnut when you're on a diet, or not drink another glass of wine at the office holiday party. It's far from clear how these examples of psychological inhibition are related to inhibitory synapses, but it's at least plausible that there's some sort of connection.

The need for inhibition might be the chief reason why the brain relies so heavily on synapses that transmit chemical signals. There is actually another kind of synapse, one that directly transmits electrical signals without using neurotransmitter. Such electrical synapses work more quickly, since they eliminate the time-consuming steps of converting signals from electrical to chemical and then back to electrical, but there are no inhibitory electrical synapses, only excitatory ones. Perhaps because of this and other limitations, electrical synapses are much less common than chemical ones.

Given that inhibition is a factor, how should our voting model be revised? Earlier I mentioned that a neuron spikes when the number of "yes" votes exceeds a threshold. If we include inhibition, spiking happens when "yes" votes exceed "no" votes by some margin set by the threshold. Like their excitatory brethren, inhibitory synapses can be stronger or weaker, so the vote is weighted rather than totally democratic. Some inhibitory synapses are even strong enough to effectively veto many excitatory synapses.

There's one last thing to know about neural voting. Neurons behave like conformists or contrarians, because they too can be classified as either excitatory or inhibitory. An excitatory neuron makes only excitatory synapses on other neurons, while an inhibitory neuron makes only inhibitory synapses. A similar uniformity does not hold for the synapses *received* by a neuron, which can be a mixture of excitatory and inhibitory.

In other words, an excitatory neuron either says "yes" to all other neurons by spiking or abstains by remaining silent. Similarly, an inhibitory neuron chooses between "no" and abstaining. A neuron cannot say "yes" to some neurons and "no" to others, or "yes" at some times and "no" at others.

If an excitatory neuron hears many "yes" votes, it also *says* "yes," conforming to the crowd. If an inhibitory neuron hears many "yes" votes, it says "no," bucking the trend. In many brain regions, including the cortex, most neurons are excitatory. You could think of the brain as being like our society, which abounds in conformists but also harbors some contrarians.

Certain sedatives work by increasing the strength of inhibition, empowering the inhibitory neurons to dampen activity. And drugs that weaken inhibition give the upper hand to excitatory neurons, which may go out of control and ignite epileptic seizures. Here you could think of excitatory neurons as rabble rousers who incite the mob to riot, whereas inhibitory neurons are like the police, summoned to dampen the excitement of the crowd.

Many other properties of synapses are under investigation by neuroscientists. But I hope it's clear that saying two neurons are "connected" only begins to describe their interaction. The connection may occur through one or more synapses — chemical or electrical or both. A chemical synapse has a direction, may be excitatory or inhibitory, and may be strong or weak. The electrical currents it produces may be lengthy or brief. All of these factors matter when synapses cause neurons to spike.

I've explained that neural pathways diverge from the eye to both the legs and the salivary glands. To make clear why any given stimulus activates some pathways but not others, I've focused on synaptic convergence, which is crucial for spiking by the voting model. If a neuron doesn't spike, it functions as a dead end for all the pathways converging onto it. The myriad dead ends imposed by nonspiking neurons are essential for brain function. They allow the sight of a snake to *not* trigger the salivary glands, and the sight of a steak to *not* make you run away.

Failing to spike is just as important to neural function as spiking. That's why single synapses and single pathways are not capable of relaying spikes. In the voting model, there are two mechanisms for making neurons choosy about when to spike. I mentioned that the axon spikes only when the total electrical current collected by the cell body exceeds some threshold. Raising the threshold for an axon is a way of making the neuron even choosier. If a neuron receives a "no" vote from an inhibitory synapse, that also increases its selectivity, as now even more "yes" votes are required for a spike. In other words, there are two mechanisms that prevent neurons from spiking indiscriminately: the threshold for spiking and synaptic inhibition.

Spikes have two functions. The generation of a spike near the cell body represents the making of a decision. The propagation of a spike along the axon communicates the result of the decision to other neurons. Communication and decision-making have different goals. The goal of communication is to preserve information, to transmit it without change. But discarding information is fundamental to making decisions. Imagine a friend trying on a coat in a boutique, unable to decide whether to purchase it. There are many inputs to his or her decision, such as the color, the fit, the designer label, the ambiance of the store, and so on. You might listen to your friend go on and on about this information. But at some point you'll lose patience and ask, "Are you buying this coat or not?" In the end, the final decision — not the many reasons for it — is what matters.

Likewise, an outgoing spike indicates that a neuron's tally of votes exceeded its threshold, but does not convey details about the individual votes of its "advisors." So neurons may transmit some information, but they also throw a lot away. (I'm reminded of my father, who likes to say proudly, "Do you know why I'm so smart? It's because I'm so good at forgetting the right things.") That's why the brain is far more sophisticated than a telecom network. It would be appropriate to say that neurons *compute,* not just communicate. We've come to associate the notion of computation exclusively with our desktop and laptop computers, but these are just one type of computational device. The brain is another — albeit a very different kind.

Though we should be cautious about comparing brains to computers, they are similar in at least one important respect. They are both "smarter" than the elements from which they're constructed. According to the weighted voting model, neurons perform a simple operation, one that does not require intelligence and can be performed by a basic machine.

How could brains be so sophisticated when neurons are so simple? Well, maybe a neuron is not so simple; real neurons are known to deviate somewhat from the voting model. Nevertheless, a single neuron falls far short of being intelligent or conscious, and somehow a network of neurons is.

This idea might have been difficult to accept centuries ago, but now we've become accustomed to the idea that an assembly of dumb components can be smart. None of the parts in a computer is by itself capable of playing chess — but a huge number of these parts, when organized in the right way, can collectively defeat the world champion. Similarly, it's the organized operation of your billions of dumb neurons that makes you smart. This is the deepest question of neuroscience: How could the neurons of your brain be organized to perceive, think, and carry out other mental feats? The answer lies in the connectome.

FOUR

Neurons All the Way Down

SPIKES AND SECRETIONS. Is there really nothing more to your mind than these physical events inside your brain? Neuroscientists take it for granted that there is not, but most people I've encountered resist the idea. Even neuroscience fans, who may start by peppering me with questions about the brain, often end up expressing the belief that the mind ultimately depends on some nonmaterial entity like the soul.

I don't know of any objective, scientific evidence for the soul. Why do people believe in it? I doubt that religion is the only reason. Everyone, religious or not, feels that he or she is a single, unified entity that perceives, decides, and acts. The statement "*I* saw a snake, and *I* ran away" assumes the existence of that entity. Your subjective feeling — and mine — is "I am one." In contrast, neuroscience contends that the unity of the mind is but an illusion hiding the spikes and secretions of a staggering number of neurons, a concept of the self that could be summed up as "I am many."

Which is the ultimate reality — the many neurons or the one soul?

In 1695 the German philosopher and mathematician Gottfried Wilhelm Leibniz argued for the latter:

> Furthermore, by means of the soul or form, there is a true unity which corresponds to what is called the *I* in us; such a thing could not occur in artificial machines, nor in the simple mass of matter, however organized it may be.

In the last years of his life, he took the argument one step further, asserting that machines were fundamentally incapable of perception:

> One is obliged to admit that perception and what depends upon it is *inexplicable on mechanical principles*, that is, by figures and motions. In imagining that there is a machine whose construction would enable it to think, to sense, and to have perception, one could conceive it enlarged while retaining the same proportions, so that one could enter into it, just like into a windmill. Supposing this, one should, when visiting within it, find only parts pushing one another, and never anything by which to explain a perception.

Leibniz could only imagine observing the parts of a machine that perceives and thinks — and he did so purely for the sake of arguing that no such machine could ever exist. But his fantasy has literally come true, if you regard the brain as a machine constructed from neuronal parts. Neuroscientists regularly measure the spiking of neurons in living, functioning brains. (The technology for measuring secretions is less advanced.)

Most of these measurements are done on animals, but occasionally they are performed on humans. The neurosurgeon Itzhak Fried operates on patients with severe cases of epilepsy. Like Penfield, he uses electrodes to map the brain before surgery, and also to make scientific observations (always with the consent of his patients). In a collaborative experiment with the neuroscientist Christof Koch and others, Fried showed a collection of photos to several patients and recorded neural activity in the medial part of the temporal lobe, or MTL. (*Medial* means "close to the plane dividing the left and right hemispheres.") Many neurons were studied, but one in particular became famous. Fried stumbled on a neuron that generated many spikes when a patient viewed photos of the actress Jennifer Aniston. The neuron generated few or no spikes when the patient viewed photos of other celebrities, nonfamous people, landmarks, animals, and other objects. Even a photo of Julia Roberts, another famously beautiful actress, elicited no response.

Reporters ate up the story, joking that scientists had finally identified the neurons in our brain that store useless information. They made quips like "Angelina Jolie may have gotten Brad Pitt, but Jennifer Aniston is the one with her own namesake neuron." They gleefully noted that the neuron remained quiet when presented with photos of Jennifer Aniston with the actor Brad Pitt. (The paper by Fried and his collaborators appeared in 2005, the same year that the celebrity supercouple divorced.)

All joking aside, how should we think about this neuron? Before drawing any conclusions, you should know that other neurons were studied too. There was a "Julia Roberts neuron" that spiked only for photos of Julia Roberts, a "Halle Berry neuron," a "Kobe Bryant neuron," and so on. Based on these findings, we could venture a theory: For every celebrity you know, there exists a "celebrity neuron" in your MTL—a neuron that spikes in response to that particular celebrity.

To be even bolder, we might suggest that this is the way perception works more broadly. This general ability is too complex to be carried out by a single neuron. Instead, it is divided up into many specific functions, each of which is the detection of some person or object and is carried out by a corresponding neuron. You might compare the brain to an army of paparazzi employed by a magazine that seeks to publish titillating photos of movie stars. Each photographer is assigned to a single celebrity. One hounds Jennifer Aniston with his camera, another devotes himself to Halle Berry, and so on. Every week, their activities determine which celebrities appear in the magazine, just as the spiking of MTL neurons determines which celebrities are perceived by a person.

Have we refuted Leibniz? It seems that we've just peeked inside the machine and seen perception reduced to spikes. But let's pause for a moment of caution. Although Fried's experiment is fascinating, it had a major limitation: Relatively few celebrities were studied. Overall, each patient viewed photos of only ten or twenty celebrities. We can't exclude the possibility that the "Jennifer Aniston neuron" would have been activated if a photograph of some other celebrity had been shown.

So let's revise our theory a bit. In our preliminary theory, we assumed a one-to-one correspondence between neurons and celebrities. Suppose instead that a neuron responds to a small percentage of celebrities, rather than only one. And suppose that each celebrity activates a small percentage of neurons, rather than just one. The spiking of this *group* of neurons is the event in the brain that marks the perception of that celebrity. (The groups activated by different celebrities are allowed to overlap partially but not completely. You can imagine that each photographer in our army of paparazzi would be assigned to cover more than one celebrity, and each celebrity would be hounded by a group of photographers.)

You might protest that perceptions are too complex to be reduced to something as simple as spiking. But remember that the spiking of a *population* of neurons defines a pattern of activity in which some neurons spike and others do not. The number of possible patterns is huge — more than enough to uniquely represent every celebrity, and indeed every possible perception.

So Leibniz was wrong. Observing the parts of the neuronal machine has told us a great deal about perception, even though neuroscientists have generally been limited to measuring spikes from a single neuron at a time. Some have measured spikes from tens of neurons simultaneously, but even this is meager compared with the enormous number of neurons in the brain. From the experiments that have been done so far, we might extrapolate: If I could observe the activities of *all* your neurons, I would be able to decode what you are perceiving or thinking. This kind of mind reading would require knowing the "neural code," which you can picture as a huge dictionary. Each entry of the dictionary lists a distinct perception and its corresponding pattern of neural activity. In principle, we could compile this dictionary by recording the activity patterns generated by a huge number of stimuli.

Physicist, mathematician, astronomer, alchemist, theologian, and Master of the Royal Mint — Sir Isaac Newton pursued many careers in a single lifetime. He invented calculus, a branch of mathematics essential to the physical sciences and engineering. He explained how planets orbit around the sun by applying his famous Three Laws of Motion and the Universal Law of Gravitation. He theorized that light is composed of particles, and discovered mathematical laws of optics describing how the paths of these particles are bent by water or glass to produce the colors of the rainbow. During his lifetime Newton was already recognized as a transcendent genius. When he died in 1727, the English poet Alexander Pope composed the epitaph: "Nature and nature's laws lay hid in night; / God said 'Let Newton be' and all was light." In a 2005 poll conducted by England's Royal Society, Isaac Newton was voted even greater than Albert Einstein.

We exalt the lone genius through such comparisons and through honors like the Nobel Prize. But another view of science places less emphasis on the individual. Newton himself acknowledged his intellectual debts by writing, "If I have seen further it is only by standing on the shoulders of giants."

Was Newton really so special? Or did he just happen to be in the right place at the right time and put two and two together? Calculus was independently invented around the same time by Leibniz. Stories like this — of nearly simultaneous discovery — are common in the history of science, because new ideas are created by combining old ideas in a new way. At any given moment in history, more than one scientist could potentially find the right combination. Since no idea is truly new, no scientist is truly special. We cannot understand the accomplishments of one without knowing how she or he drew on the ideas of others.

Neurons are like scientists in this regard. If a neuron spikes in response to Jennifer Aniston but not other celebrities, we might think that the neuron's function is the detection of Jen. But this neuron is embedded in a network of many other neurons. It would be a mistake to think of this neuron as a lone genius, detecting Jen all by itself. Newton's words ring even truer for neurons than for Newton: "If a neuron sees further, it is only by standing on the shoulders of other neurons." To understand how a neuron manages to detect Jen, we need to know something about the neurons from which it receives information.

The weighted voting model I presented earlier forms the basis for a theory of what happens. Let's describe Jen as a combination of simpler parts. She has blue eyes, blond hair, an angular chin, and so on (as of this writing, anyway). If the list is long enough, it will uniquely describe Jen and no other celebrity. Now suppose that the brain contains neurons for detecting each stimulus in the list. There is a "blue-eye neuron," a "blond-hair neuron," and an "angular-chin neuron." Now here is the central hypothesis: The "Jennifer Aniston neuron" receives excitatory synapses from all of these "part neurons." The threshold of the "Jennifer Aniston neuron" is high, so it spikes only when all of the part neurons spike, a unanimous vote that happens only in response to Jen. In short, a neuron detects Jen as a combination of Jen parts, which are detected by other neurons.

This explanation sounds reasonable, but it raises more questions. How does the "blue-eye neuron" manage to detect blue eyes, the "blond-hair neuron" detect blond hair, and so on? I'm reminded of the funny story that opens the book *A Brief History of Time* by the physicist Stephen Hawking:

A well-known scientist . . . once gave a public lecture on astronomy. He described how the earth orbits around the sun and how the sun, in turn, orbits around the center of a vast collection of stars called our galaxy. At the end of the lecture, a little old lady at the back of the room got up and said: "What you have told us is rubbish. The world is really a flat plate supported on the back of a giant tortoise." The scientist gave a superior smile before replying, "What is the tortoise standing on?" "You're very clever, young man, very clever," said the old lady. "But it's turtles all the way down!"

Likewise, my answer is "It's neurons all the way down." A blue eye is a combination of simpler parts: a black pupil, a blue iris, a white area surrounding the iris, and so on. Therefore a "blue-eye neuron" can be constructed by wiring it to neurons that detect these parts of a blue eye. Unlike the old lady, I can avoid the problem of infinite regress. If we keep on dividing each stimulus into a combination of simpler parts, eventually we will end up with stimuli that cannot be divided further: tiny spots of light. Each photoreceptor in the eye detects a tiny spot of light at a particular location in the retina. There is little mystery in that. Photoreceptors are similar to the many tiny sensors in your everyday digital camera, each of which detects the light at a single image pixel.

According to this theory of perception, neurons are wired into a network with a hierarchical organization. Those at the bottom detect simple stimuli like spots of light. As we ascend the hierarchy, neurons detect progressively more complex stimuli. Neurons at the top detect the most complex stimuli, such as Jennifer Aniston. The wiring of the network obeys the following rule:

> *A neuron that detects a whole receives excitatory synapses from neurons that detect its parts.*

In 1980 the Japanese computer scientist Kunihiko Fukushima simulated an artificial neural network for visual perception, which was wired up with a hierarchical organization governed by this rule. His Neocognitron network was a descendant of the *perceptron* introduced by the American computer scientist Frank Rosenblatt in the 1950s. A perceptron contains layers of neurons "standing on the shoulders" of other neurons, as shown in Figure 19. Each neuron receives connections only from neurons in the layer just below.

The Neocognitron recognized handwritten characters. Its descendants display more impressive visual capabilities, such as recognizing

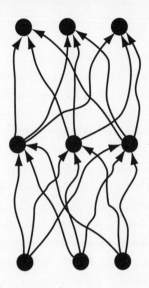

FIGURE 19. *A multilayer perceptron model of a neural network*

objects from photographs. Although these artificial neural networks still make more mistakes than human beings do, their performance is improving year after year. This engineering success lends some plausibility to the *hierarchical perceptron* model for the brain.

In the wiring rule introduced above, we focused on how a neuron receives synapses from neurons that are lower in the hierarchy. Alternatively, we can look in the opposite direction and specify how a neuron sends synapses to neurons higher in the hierarchy:

> *A neuron that detects a part sends excitatory synapses to neurons that detect its wholes.*

The two formulations of the rule are equivalent, because a stimulus detected by a neuron somewhere in the middle of the hierarchy can be regarded either as a whole containing a number of simpler parts, or as a part that belongs to a number of more complex wholes. Again taking a blue eye as our example of a stimulus, we can see it as containing simpler parts like the pupil, the iris, and the white, or as being part of more complex wholes like Jennifer Aniston, Leonardo DiCaprio, and the many other people who have blue eyes.

So the function of a neuron depends on its output connections, not only its input connections. To clarify this counterpoint, let's embellish the story of Newton and Leibniz. Suppose you read in the news about the unearthing of old documents proving that some unknown mathematician invented calculus fifty years before Newton and Leibniz did. After failing to convince others to pay attention, she died in obscurity and took calculus to her grave. Should we now rewrite the history books, crediting this unsung scholar rather than Newton and Leibniz?

Such revisionist history might sound fairer, but it would fail to recognize the social aspect of science. Earlier I argued that discovery is not just an individual creative act of a lone genius, because any new idea depends on old ideas borrowed from other people. In the same vein, one might argue that the act of discovery includes not only the creation of a new idea but also the act of persuading others to accept it. To receive full credit for a discovery, a person must influence others.

Newton's place in history is defined by how he used the ideas of his predecessors and shaped the ideas of his successors. Similarly, I'd like to propose that:

> The function of a neuron is defined chiefly by its connections with other neurons.

This mantra defines a doctrine I'll call *connectionism*. It encompasses both input and output connections. To know what a neuron does, we must look at its inputs. To understand the effects of a neuron, we should look at its outputs. Both of these perspectives were taken above in our two formulations of the part–whole rule of wiring introduced for perception. As we continue our exploration of connectionist theories, we'll encounter plausible explanations of memory and other mental phenomena, in addition to perception.

That sounds fascinating, but is there any solid evidence for these theories in real brains? Unfortunately, we've lacked the right experimental techniques to find out. In the case of perception, neuroscientists haven't been in a position to find the neurons wired to the Jennifer Aniston neuron, and to see whether they indeed detect Jen parts. More generally, if we accept the defining mantra of connectionism, it follows that we cannot truly understand the brain without mapping neural connections — in other words, finding connectomes.

Here's a wonderful thing about the brain: You can think about Jennifer Aniston even if you are not watching her on television or seeing her in a magazine. Thinking of Jen does not require *perceiving* her; you are thinking of her if you recall her performance in the 2003 film *Bruce Almighty*, fantasize about meeting her, or contemplate her latest love interest. Can thinking, like perception, also be reduced to spikes and secretions?

Let's return to the experiment of Itzhak Fried and his collaborators for some clues. Their "Halle Berry neuron" was activated by an image of the actress Halle Berry, suggesting that it plays a role in perceiving her. But the neuron was also activated by the written words *Halle Berry*, indicating that it participates in thinking about her as well. So it seems that the "Halle Berry neuron" represents the abstract *idea* of Halle Berry, which can arise from either perception or thought.

Both phenomena can be regarded as specific examples of a more general operation: association. Perception is the association of an idea with a stimulus, while thought is the association of an idea with another idea. So how do perception and thought work together when you're recalling a memory? Let's consider a scenario.

It's a fine spring morning, and you are walking down the street on the way to work. You catch the scent of flowers; within a few steps the smell becomes overpowering. You're not yet conscious of the magnolias blooming at the side of the road, but all of a sudden you're transported far away. You remember standing next to a magnolia tree, outside the red brick house of your first sweetheart. He is holding you in his arms. You feel shy and embarrassed. A plane is flying overhead, and you hear his mother calling for you to come have a glass of lemonade.

By the time the recollection is complete, you are thinking of many ideas: the magnolia, the red brick house, your sweetheart, the plane, and so on. For each of these ideas, let's suppose there exists a corresponding neuron in your brain. A "magnolia neuron," a "red brick house neuron," a "sweetheart neuron," a "plane neuron" — all are spiking as you recollect your first kiss.

How was all this spiking triggered by the magnolia smell? The spiking of the "magnolia neuron" was caused by neural pathways from your nose. But how can we explain why the "plane neuron" is active even though there is no plane in the sky, and why the "red brick house neuron" is active even though there is no red brick house? This must be the result of thinking, not perception.

To explain all this activity, let's hypothesize that the neurons are excitatory and are mutually connected by synapses into a structure known as a *cell assembly*. The one shown in Figure 20 is just a small example, but you could imagine a larger assembly containing many neurons all connected with each other. Omitted from the diagram are connections to and from other neurons in the brain. These connections would bring signals from sense organs or send signals to muscles. Here we focus on the connections within the cell assembly, which represent the associations involved in thought.

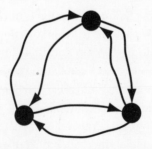

FIGURE 20. *A cell assembly*

How do these connections trigger the recollection of your first kiss? Since the neurons are assumed excitatory, the activation of the "magnolia neuron" excites the other neurons in the cell assembly to become active. You can imagine it like a forest fire jumping from tree to tree, or a flash flood surging through a web of desert ravines. A similar spreading of neural activity allows the magnolia smell to trigger the recollection of all the ideas involved in the entire memory of your first kiss.

Memory is wonderful when it works, but we've all noticed and complained about its failures as well. In fact, a feeling of difficulty often accompanies the experience of memory, while perception usually feels effortless. If the brain stored only a single memory in a single cell assembly, perhaps remembering would be a trivial task too. But many assemblies are required to store many memories. If cell assemblies were like islands, completely independent of each other, having many of them would be no problem. But it turns out they need to overlap, and that's where the possibility of failure creeps in.

Recall that the memory of your first kiss included your sweetheart's mother calling for you to have a glass of lemonade. Let's say you have another memory involving lemonade, from the hot summer day when you sat in front of your house and sold ice-cold lemonade in paper cups to passersby. This memory is different from that of your first kiss, but they have lemonade in common, so their cell assemblies overlap in the "lemonade neuron," as shown in Figure 21. (The double-headed arrows represent synapses going in both directions.) The danger of

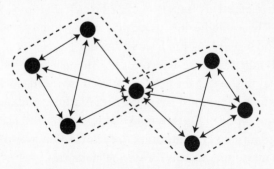

FIGURE 21. *Overlapping cell assemblies*

overlap is obvious: Activating one of these cell assemblies might also ignite the other. The magnolia smell might activate a mishmash of two memories, a confused combination of your first kiss and the lemonade stand. This scenario could be a cause of inaccurate memory recall more generally.

To prevent the indiscriminate spread of activity, the brain could give each neuron a high threshold for activation. Let's suppose that a neuron is not activated unless it receives at least two "yes" votes from its advisors. Since the cell assemblies of Figure 21 overlap only in a single neuron, activity will not spread from one to the other.

But the protection mechanism of a high threshold has its own pitfall. It also makes the criterion for recalling a memory more stringent. Because of it, activation of at least two neurons in a cell assembly is necessary to cause recollection of the entire memory. The magnolia smell alone would not be enough to trigger recall of your first kiss. It would have to be accompanied by the sound of a plane overhead, or some other stimulus that was part of your first kiss.

Whether the brain should be that selective about recollection depends on the details of the situation. But it's clear that activity might sometimes fail to spread even when it should. This could be the cause of another common complaint about memory, the failure to recall anything at all. (It doesn't explain the tantalizing "tip-of-your-tongue" feeling, but it could explain the failure that causes the feeling.) So I imagine the brain's memory systems as balanced on a knife edge. Too much spread of activity leads to confused recall, while too little causes no recall. This could be one reason why memory can never function perfectly, no matter how much we wish it did.

The amount of overlap between cell assemblies depends on how many we try to jam into the network. Clearly, the overlap will become large if we try to store too many memories. At some point there will no longer be any value of the threshold that both allows recollection and prevents confusion. This catastrophe of information overload sets the network's maximum capacity for storing memories.

In the cell assembly, all neurons make synapses on all other neurons, so any part of the memory can trigger recall of the rest. A photo of your sweetheart might trigger recollection of his house, and a visit to his house might trigger recollection of him. Recollection is bidirectional in this case, but there are also cases in which it has a unique direction, as in a memory that is essentially a story, a sequence of events unfolding in a particular chronological order. How do we account for that? The obvious answer is to arrange the synapses so that activity can flow in one direction. In the *synaptic chain* shown in Figure 22, activity spreads from left to right.

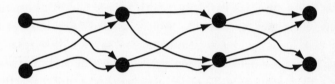

FIGURE 22. *A synaptic chain*

Let me summarize this theory of recollection. Ideas are represented by neurons, associations of ideas by connections between neurons, and a memory by a cell assembly or synaptic chain. Memory recall happens when activity spreads after ignition by a fragmentary stimulus. The connections of a cell assembly or synaptic chain are stable over time, which is how a childhood memory can persist into adulthood.

The psychological component of this theory is known as *associationism,* a school of thought that began with Aristotle and was later revived by English philosophers such as John Locke and David Hume. By the late nineteenth century, neuroscientists had recognized the existence of fibers in the brain and were speculating about pathways and connections. It was only logical to suppose that physical connections are the material basis of psychological associations.

The theory of connectionism was developed by several generations of researchers in the second half of the twentieth century. Over the decades, it was dogged by a persistent set of critiques. As early as 1951, Karl Lashley, the originator of cortical equipotentiality, had published a withering attack in his famous paper "The Problem of Serial Order in Behavior." His first critique was rather obvious: The brain can generate a seemingly infinite variety of sequences. A synaptic chain might be ideal for reciting a poem, for generating the same sequence of words every time, but doesn't seem appropriate for normal language, in which the same sentence is rarely ever repeated exactly.

This first concern of Lashley's is fairly easy to address. Imagine a synaptic chain that diverges into two chains, like a fork in the road. These two chains could diverge into four, and so on. If there are many branch points in a network, it could potentially generate a huge variety of activity sequences. The trick here is to make sure that activity always "chooses" one branch or the other, but not both. Theorists have shown this can be done through inhibitory neurons that are wired up to make the branches "compete" with each other.

Lashley's second, more fundamental critique focused on the problem of syntax. A synaptic chain uses connections to represent the association of one idea with the next in the sequence. Lashley pointed out that generating a grammatical sentence is not so simple, because "each syllable in the series has associations not only with adjacent words in the series, but also with more remote words." Whether the end of a sentence is correct may depend on the exact arrangement of words at the beginning of the sentence. Lashley's ideas prefigured the later emphasis of the linguist Noam Chomsky and his many followers on the problem of syntax.

Connectionists have also addressed Lashley's second critique, though a discussion of this research is outside the scope of this book. In any case, researchers have shown that connectionism is not as limited as its critics initially believed. I don't think it's possible to reject the doctrine on purely theoretical grounds; it needs to be tested empirically, and connectomics can be used to do that, as I'll explain later.

But first let me complete the theory. The hypothesis that synapses are the material basis of associations and that recollections arise from cell assemblies and synaptic chains is only half the story. It's time to confront a question I've postponed until now: How is a memory stored in the first place?

FIVE

The Assembly of Memories

THE GREAT PYRAMID of Giza has stood for forty-five hundred years, an island of eternity in the shifting desert sands near Cairo. Its massive form invites awe, but just one of its large blocks is imposing enough. No one knows for sure how the two-and-a-half-ton stones were cut at the quarry, transported to the site, and lifted up to 140 meters off the ground. If construction took twenty years, as the ancient Greek historian Herodotus estimated, the 2.3 million blocks were placed at the staggering rate of one every minute.

The Egyptian pharaoh Khufu built the Great Pyramid to serve as his tomb. If we were not separated from the suffering of one hundred thousand workers by the cool distance of history, we might condemn the pyramid as a cruel display of power by an egotistical despot. But perhaps it is better to forgive Khufu and simply marvel at the fantastic accomplishment of these nameless workers. We can regard the pyramid not as a monument to the pharaoh but as a testament to human ingenuity.

Khufu's strategy was straightforward: If you want to be remembered forever, build a massive structure out of material durable enough to survive the ravages of time. By the same token, perhaps the brain's ability to remember depends on the persistence of its material structure. What else could account for the indelibility of memories that last an entire lifetime? Then again, we sometimes forget or misremember, and we add new memories every day. That's why Plato compared memory to another kind of material, one more flexible than the pyramid's stone blocks:

> There exists in the mind of man a block of wax. . . . Let us say that this tablet is a gift of Memory, the mother of the Muses; and that when we wish to remember anything . . . we hold the wax to the perceptions and thoughts, and in that material receive the impression of them as from the seal of a ring.

In the ancient world, wooden boards coated with wax were a common sight, functioning much like our modern-day notepads. A sharp stylus was used to write text or draw diagrams in the wax. Afterward, a straight-edged instrument smoothed the wax, erasing the tablet for its next use. As an artificial memory device the wax tablet served as a natural metaphor for human memory.

Plato did not mean, of course, that your skull is literally filled with wax. He imagined some analogue—a material that could hold its shape and could also be reshaped. Artisans and engineers mold "plastic" materials and hammer or press "malleable" ones. Likewise, we say that parents and teachers mold young minds. Could that be more than metaphor? What if education and other experiences literally reshape the material structure of the brain? People often say that the brain is plastic or malleable, but what exactly does this mean?

Neuroscientists have long hypothesized that the connectome is the analogue of Plato's wax tablet. Neural connections are material structures, as we've seen from electron microscope images. Like wax, they are stable enough to remain the same for long periods of time, but they are also plastic enough to change.

One important property of a synapse is its strength, its weight in the vote conducted by a neuron when "deciding" when to spike. It's known that synapses can strengthen and weaken; you can think of such changes as *reweighting.* What exactly happens at a synapse when it strengthens? The discoveries of the many neuroscientists who are investigating this question could fill an entire book. Here I'll only give a simplistic answer, one that the phrenologists would have liked: Synapses strengthen by getting bigger. Recall that there are neurotransmitter vesicles on one side of the synaptic cleft, and neurotransmitter receptors on the other. A synapse strengthens by creating more of both. To release more neurotransmitter in each secretion, it amasses more vesicles. To be more sensitive to a given amount of neurotransmitter, it deploys more receptors.

Synapses can also be created and eliminated, a phenomenon I'll call *reconnection.* It has long been known that young brains create synapses in droves as neurons connect themselves into a network. The creation of a synapse happens at a point of contact between two neurons. For reasons that are not well understood, vesicles, receptors, and other types of synaptic machinery aggregate at this point. Young brains eliminate synapses as well, by removing such molecular machinery from contact points.

In the 1960s most neuroscientists believed that synapse creation and elimination ceased by adulthood. Their belief was based on theoretical preconceptions rather than empirical evidence. Maybe they thought of brain development as resembling the construction of an electronic device: We have to connect a lot of wires to build the device, but we never reconnect them differently after it becomes operational. Or maybe they thought of synapse strength as being easy to modify, like computer software, but considered the synapses themselves to be fixed like hardware.

In the last ten years neuroscientists have done an about-face. It is now widely accepted that synapses are created and eliminated even in adult brains. Convincing evidence was finally obtained directly, by watching synapses in living brains using a new imaging method known as two-photon microscopy. The images in Figure 23 show a dendrite in the cortex of a mouse changing over the course of two weeks. (The day is indicated by the number in the lower left of each image.)

The dendrite bears thornlike protuberances known as spines. Most

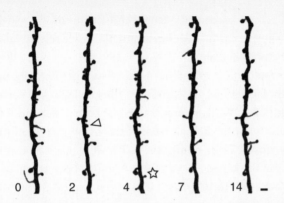

FIGURE 23. *Evidence for reconnection: spines appearing and disappearing on a dendrite in the cortex of a mouse*

synapses between excitatory neurons are made onto spines rather than onto the shaft of the dendrite. In the figure, some spines are stable for the whole two weeks, but others appear (for example, look at the spine indicated by the arrowhead) and disappear (see the starred spine). This is good evidence that synapses are being created and eliminated. Researchers still debate how frequently such reconnection happens, but all agree that it is possible.

Why are reweighting and reconnection so important? These two types of connectome change continue to happen for our entire lives. We must study them if we want to understand personal change as a lifelong phenomenon. No matter how old we get, we never stop storing new memories, barring some kind of brain disorder. As we age, we may complain that it's more difficult to learn, but even the elderly can acquire new skills. It seems likely that reweighting and reconnection are involved in such changes.

But do we have any proof? Evidence implicating reweighting in memory storage has come from Eric Kandel and his collaborators, who studied the nervous system of *Aplysia californica*, a squishy creature found in tide pools of California beaches. This animal retracts its gill and siphon when disturbed, and can become more or less sensitive to disturbances — a simple kind of memory. We previously learned that such behaviors depend on neural pathways from sense organs to muscles. Kandel identified a single connection in the relevant pathway and showed that changes in its strength were related to the simple memory mentioned above.

Is reconnection involved in memory storage? Earlier I mentioned the phrenological idea of learning as thickening of the cortex. In the 1970s and 1980s, William Greenough and other researchers found evidence that such thickening was caused by an increase in the number of synapses. Their findings — which were made by counting synapses in the thickened cortex of rats who had been raised in enriched cages — led some to propose a neo-phrenological theory: Memories are stored by creating synapses.

Neither of these approaches truly succeeded in elucidating memory storage, however. Kandel's approach has faltered for brains more like our own, in which memories do not appear localized to single synapses. It seems more probable that memories are stored as patterns of many connections. Greenough's approach is also incomplete, because counting synapses does not tell us how they are organized into patterns. Furthermore, increases in synapse number, like cortical thickening, are correlated with learning, but it's not clear whether they are causally related.

To really crack the problem of memory, we need to figure out whether reweighting and reconnection are involved, and if so, exactly how. Earlier I explained the theory that the patterns of connection relevant for memory are cell assemblies and synaptic chains. Here I'll take a further step and propose that these patterns are created by reweighting and reconnection, and I'll explore the many questions that arise. Are these two processes independent, or do they work together? Why would the brain use both rather than just one? Can we explain some limitations of memory as malfunctions of these storage processes?

Beyond satisfying our basic curiosity about memory, research on reweighting and reconnection could have practical consequences. Suppose that your goal is to develop a drug that improves memory storage. If you believe neo-phrenology, you might try to develop a drug that enhances the molecular processes involved in synapse creation. But if neo-phrenology is wrong — as it most likely is — your creation of more synapses might have effects very different from what you intended. More generally, whether we want to improve our memory abilities or prevent them from malfunctioning, knowledge about the basic mechanisms will be essential.

We've seen how a cell assembly might retain associations between ideas as connections between neurons. But how does the brain create a cell assembly in the first place? This is the connectionist version of a much older question posed by philosophers: Where do ideas and their associations come from? While some might be innate, it's clear that others must be learned from experience.

Over the ages, philosophers came up with a list of principles by which associations can be learned. At the top of the list is coincidence, sometimes called contiguity in time or place. If you see photos of a pop singer with her baseball-player boyfriend, you will learn an association between them. A second factor is repetition. Seeing these celebrities together just once might not be enough to create the association in your mind, but if you see them ad nauseam day after day in every magazine and newspaper, you will not be able to avoid learning the association. Ordering in time also seems important for some associations. As a child you recited the letters of the alphabet repeatedly until you knew them by heart. You learned the association from each letter to the next, since the letters always followed one another in the same sequence. In contrast, the association between the pop singer and her boyfriend will be bidirectional, since they always appear simultaneously.

So philosophers proposed that we learn to associate ideas when one repeatedly accompanies or succeeds another. This inspired connectionists to conjecture:

> *If two neurons are repeatedly activated* simultaneously, *then the connections between them are strengthened in both directions.*

This rule of plasticity is appropriate for learning two ideas that repeatedly occur together, like the pop singer and her boyfriend. For learning associations between sequential ideas, connectionists proposed a similar rule:

> *If two neurons are repeatedly activated* sequentially, *the connection from the first to the second is strengthened.*

In both rules, by the way, it's assumed that the strengthening is permanent or at least long-lasting, so that the association can be retained in memory.

The sequential version of the rule was hypothesized by Donald Hebb, who also proposed the cell assembly in his 1949 book, *The Organization of Behavior.* Both simultaneous and sequential versions have come to be known as Hebbian rules of synaptic plasticity. Both are said to be "activity-dependent," because plasticity is triggered by the activity of the neurons involved in the synapse. (There are other ways of inducing synaptic plasticity that do not involve activity, such as the application of certain drugs.) Typically, Hebbian plasticity refers only to synapses between excitatory neurons.

Hebb was way ahead of his time. Neuroscientists had no means of detecting synaptic plasticity. In fact, they could not even measure synaptic strengths at all. Measurements of spiking had been conducted for decades using metal wires inserted into the nervous system. Since the tip of the wire remained outside the neuron, this method was known as "extracellular" recording. The signals from the wire carried the spikes of several neurons, mixing them together like conversations in a crowded bar. This method, still in use today, is the one that was employed by Itzhak Fried and his collaborators to find the "Jennifer Aniston neuron." By carefully maneuvering the tip of the wire, it's possible to isolate the spikes of a single neuron, much as you do when you stick your ear close to the mouth of one of your friends at the bar.

While extracellular recording was sufficient for detecting spikes, it failed to measure the weak electrical effects of individual synapses. This was first accomplished in the 1950s by inserting a glass electrode with an extremely sharp tip into a single neuron. Such "intracellular" recording is so precise that it can detect signals much weaker than spikes, the equivalent of sticking your ear *inside* the mouth of a speaker at a bar. An intracellular electrode can also be used to stimulate a neuron to spike, by injecting electrical current into the neuron.

To measure the strength of a synapse from neuron A to neuron B, we insert electrodes into both neurons; we stimulate neuron A to spike, which causes the synapse to secrete neurotransmitter; and we measure the voltage of neuron B, which responds with a blip. The size of this blip is the strength of the synapse.

Along with measuring a synapse's strength, we can also measure *changes* in its strength. To induce Hebbian plasticity, we stimulate spiking in a pair of neurons. Repeated stimulation, either sequential or simultaneous, has been shown to strengthen synapses in accordance with the two versions of the Hebbian rule given earlier.

After a change in synaptic strength has been induced, it can last for the rest of the experiment — a few hours at most, as it's not easy to keep the neurons alive after they've been penetrated with electrodes. But cruder experiments involving populations of neurons and synapses, first done in the 1970s, suggest that changes in synaptic strength can last for weeks or longer. The issue of persistence is critical if Hebbian plasticity is to be the mechanism of memory storage, as some memories can last for a lifetime.

These experiments from the 1970s provided the first evidence for synaptic strengthening. By that time a theory of memory storage had also emerged, based on Hebb's original ideas. In the simplest version of the theory, a network starts out with weak synapses in both directions between the neurons of every pair. This assumption will turn out to be problematic, but let's accept it for now, for the purpose of introducing the theory.

Return to the scene of your first kiss, the actual event that imprinted your memory. The "magnolia neuron," the "red brick house neuron," the "sweetheart neuron," the "plane neuron," and so on were being activated by the stimuli around you — quite vigorously, I imagine. If we assume the simultaneous version of the Hebbian rule, all this spiking strengthened the synapses between these neurons.

The strengthened synapses together constitute a cell assembly, if we redefine this concept to mean a set of excitatory neurons mutually interconnected by *strong* synapses. Our original definition didn't have this stipulation. We need it now because the network contains many weak synapses that do not belong to the cell assembly. They existed before your first kiss, and remained unchanged afterward.

The weak synapses have no effect on recollection. Activity spreads from neuron to neuron within the cell assembly but does not spread any farther, because synapses from the cell assembly to other neurons are too weak to activate them. Thus the new definition of a cell assembly functions just as the old one did.

An analogous theory applies for the synaptic chain. Suppose that a sequence of stimuli activates a sequence of ideas. Each idea is represented by the spiking of a group of neurons. If the groups spike in this sequence repeatedly, the sequential version of the Hebbian rule will strengthen all existing synapses from neurons in each group to neurons in the next group. This is a synaptic chain, if we redefine this concept to mean a pattern of *strong* connections.

If the connections are sufficiently strong, then the spiking will propagate through the chain without any need for a sequence of external stimuli. Any stimulus that activates the first group of neurons will trigger the recollection of a sequence of ideas, as described in Chapter 4. Every successive recollection of the sequence will further strengthen the connections of the chain by Hebbian plasticity. This is analogous to the way that the flowing water of a stream slowly deepens its bed, making it even easier for the water to flow.

While it's important to remember things, it's also vital to forget. At one time your Jennifer Aniston and Brad Pitt neurons were linked by strong synapses into a cell assembly. But one day you started to see Brad with Angelina. (I know it was sad, but I hope you didn't feel *too* devastated.) Hebbian plasticity strengthened the connections between your Brad and Angelina neurons, creating a new cell assembly. What happened to the connections between your Brad and Jen neurons?

You could imagine an analogue of the Hebbian rule serving the function of forgetting. Perhaps the connections between two neurons are weakened if one is repeatedly active while the other is inactive. This would weaken the synapses between Brad and Jen every time you saw him without her.

Alternatively, one can imagine that weakening is caused by direct competition between synapses. Perhaps the synapses between Brad and Angelina directly compete with those between Brad and Jen for some foodlike substance that synapses need in order to survive. If some synapses strengthen, they consume more of the substance, leaving less for the others, which grow weak. It's not clear whether such substances exist for synapses, but analogous "trophic factors" are known to exist for neurons. Nerve growth factor is one example; its discovery won Rita Levi-Montalcini and Stanley Cohen a 1986 Nobel Prize.

The Romans used the phrase *tabula rasa* to refer to the wax tablets mentioned by Plato. It's traditionally translated as "blank slate," since little chalkboards replaced wax tablets in the eighteenth and nineteenth centuries. In "An Essay Concerning Human Understanding," the associationist philosopher John Locke resorted to yet another metaphor:

> Let us then suppose the mind to be, as we say, white paper, void of all characters, without any ideas. How comes it to be furnished? Whence comes it by that vast store which the busy and boundless fancy of man has painted on it with an almost endless variety? Whence has it all the materials of reason and knowledge? To this I answer, in one word, from EXPERIENCE.

A sheet of white paper contains zero information but unlimited potential. Locke argued that the mind of a newborn baby is like white paper, ready to be written on by experience. In our theory of memory storage, we assumed that all neurons started out connected to all other neurons. The synapses were weak, ready to be "written on" by Hebbian strengthening. Since all possible connections existed, any cell assembly could be created. The network had unlimited potential, like Locke's white paper.

Unfortunately for the theory, the assumption of all-to-all connectivity is flagrantly wrong. The brain is actually at the opposite extreme of *sparse* connectivity. Only a tiny fraction of all possible connections actually exist. A typical neuron is estimated to have tens of thousands of synapses, much less than the total of 100 billion neurons in the brain. There's a very good reason for this: Synapses take up space, as do the neurites they connect. If every neuron were connected to every other neuron, your brain would swell in volume to a fantastic size.

So the brain has to make do with a limited number of connections. This could present a serious problem when you are learning associations. What if your Brad and Angelina neurons had not been connected at all? When you started seeing them together, Hebbian plasticity could not have succeeded in linking the neurons into a cell assembly. There is no potential to learn an association unless the right connections already exist.

Especially if you think a lot about Brad and Angelina, it's likely that each is represented by many neurons in your brain, rather than just one. (In Chapter 4 I argued that this "small percentage" model is more plausible than the "one and only" model.) With so many neurons available, it's likely that a few of your Brad neurons happen to be connected to a few of your Angelina neurons. That might be enough to create a cell assembly in which activity can spread from Brad neurons to Angelina neurons during recollection, or vice versa. In other words, if every idea is redundantly represented by many neurons, Hebbian learning can work in spite of sparse connectivity.

Similarly, a synaptic chain can be created by Hebbian plasticity even if some connections are missing. Imagine removing the connection represented by the dashed arrow shown in Figure 24. This would break some pathways, but there would still be others extending from the beginning to the end, so the synaptic chain could still function. Each idea in the sequence is represented by only two neurons in the diagram, but adding more neurons would make the chain even more able to withstand missing connections. Again, a redundant representation enables learning to establish associations in spite of sparse connectivity.

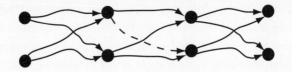

FIGURE 24. *Elimination of a redundant connection in a synaptic chain*

The ancients already knew the paradoxical fact that remembering *more* information is often easier than remembering less. Orators and poets exploited this fact in a mnemonic technique called the method of loci. To memorize a list of items, they imagined walking through a series of rooms in a house and finding each item in a different room. The method may have worked by increasing the redundancy of each item's representation.

So sparse connectivity could be a major reason why we have difficulty memorizing information. Because the required connections don't exist, Hebbian plasticity can't store the information. Redundancy solves this problem somewhat, but could there be some other solution?

Why not create new synapses "on demand," whenever a new memory needs to be stored? We could imagine a variant of Hebb's rule of plasticity: "If neurons are repeatedly activated simultaneously, then new connections are created between them." Indeed this rule would create cell assemblies, but it conflicts with a basic fact about neurons: There is negligible crosstalk between electrical signals in different neurites. Let's consider a pair of neurons that contact each other without a synapse. They could create one, but it's implausible that this event could be triggered by simultaneous activity. Because there is no synapse, the neurons can't "hear" each other or "know" they are spiking simultaneously. By similar arguments, the "on-demand" theory of creation doesn't seem plausible for synaptic chains either.

So let's consider another possibility: Perhaps synapse creation is a *random* process. Recall that neurons are connected to only a subset of the neurons that they contact. Perhaps every now and then a neuron randomly chooses a new partner from its neighbors and creates a synapse. This may seem counterintuitive, but think about the process of making friends. Before you speak with someone, it's almost impossible to know whether you should be friends. The initial encounter might as well be random — at a cocktail party, in the gym, or even on the street. Once you start to talk, you develop a sense of whether your relationship could strengthen into friendship. This process isn't random, as it depends on compatibility. In my experience, people with the richest sets of friends are open to chance meetings but also very skilled at recognizing new people with whom they "click." The random and unpredictable nature of friendship is a large part of its magic.

Similarly, the random creation of synapses allows new pairs of neurons to "talk" with each other. Some pairs turn out to be "compatible," because they are activated simultaneously or sequentially as the brain attempts to store memories. Their synapses are strengthened by Hebbian plasticity to create cell assemblies or synaptic chains. In this way, the synapses for learning an association can be created even if they don't initially exist. We may eventually succeed at learning after failing at first, because our brains are continually gaining new potential to learn.

Synapse creation alone, however, would eventually lead to a network that is wasteful. In order to economize, our brains would need to eliminate the new synapses that aren't used for learning. Perhaps these synapses first become weaker by the mechanisms discussed earlier (recall what happens when you are unlearning the Brad–Jen connection), and the weakening eventually causes the synapses to be eliminated.

You could think of this as a kind of "survival of the fittest" for synapses. Those involved in memories are the "fittest," and get stronger. Those not involved get weaker, and are finally eliminated. New synapses are continually created to replenish the supply, so that the overall number stays constant. Versions of this theory, known as neural Darwinism, have been developed by a number of researchers, including Gerald Edelman and Jean-Pierre Changeux.

The theory argues that learning is analogous to evolution. Over time, a species changes in ways that might seem intelligently designed by God. But Darwin argued that changes are actually generated randomly. We end up noticing only the good changes, because the bad ones are eliminated by natural selection, the "survival of the fittest." Similarly, if neural Darwinism is correct, it might seem that synapses are "intelligently" created, that they are generated "on demand" only if needed for cell assemblies or synaptic chains. But in fact synapses are created randomly, and then the unnecessary ones are eliminated.

In other words, synapse creation is a "dumb," random process that endows the brain only with the *potential* for learning. By itself, the process is not learning, contrary to the neo-phrenological theory mentioned earlier. This is why a drug that increases synapse creation might be ineffective for improving memorization, unless the brain also succeeds at eliminating the larger number of unnecessary synapses.

Neural Darwinism is still speculative. The most extensive studies of synapse elimination are by Jeff Lichtman, who has focused on the synapses from nerves to muscles. Early in development, connectivity starts out indiscriminate, with each fiber in a muscle receiving synapses from many axons. Over time, synapses are eliminated until each fiber receives synapses from just a single axon. In this case, synapse elimination refines connectivity, making it much more specific. Motivated to see this phenomenon more clearly, Lichtman has become a major proponent of superior imaging technologies — a topic I'll return to in later chapters.

Through the images of dendritic spines shown earlier in Figure 23, we saw that reconnection has also been studied in the cortex. The researchers showed that most new spines disappear within a few days, but a larger fraction survive when the mouse is placed in an enriched cage like the ones Rosenzweig used. Both of these observations are consistent with the idea of "survival of the fittest," that new synapses survive only if they are used to store memories. The evidence is far from conclusive, however. It's an important challenge for connectomics to reveal the exact conditions under which a new synapse survives or is eliminated.

• • •

We've seen that the brain may fail to store memories if the required connections don't exist. That means reweighting has limited capacity for storing information in connectivity that is fixed and sparse. Neural Darwinism proposes that the brain gets around this problem by randomly creating new synapses to continually renew its potential for learning, while eliminating the synapses that aren't useful. Reconnection and reweighting are not independent processes; they interact with each other. New synapses provide the substrate for Hebbian strengthening, and elimination is triggered by progressive weakening. Reconnection provides added capacity for information storage, compared with reweighting alone.

A further advantage of reconnection is that it may stabilize memories. For a clearer understanding of stability it's helpful to broaden the discussion. So far I've focused on the idea that synapses retain memories. I should mention, however, that there is evidence for another retention mechanism based on spiking. Suppose that Jennifer Aniston is represented not by a single neuron but by a group of neurons organized into a cell assembly. Once the stimulus of Jen causes these neurons to spike, they can continue to excite each other through their synapses. The spiking of the cell assembly is self-sustaining, persisting even after the stimulus is gone. The Spanish neuroscientist Rafael Lorente de Nó called this "reverberating activity," because of its similarity to a sound that persists by echoing in a canyon or cathedral. Persistent spiking could explain how you can remember what you have just seen.

Judging from many experiments, such persistent spiking appears to retain information over time periods of seconds. There is good evidence, however, that retention of memories over long periods does not require neural activity. Some victims of drowning in icy water have been resuscitated after being effectively dead for tens of minutes. Even though their hearts had stopped pumping blood, the icy cold prevented permanent brain damage. The lucky ones recovered with little or no memory loss, despite the complete inactivity of their neurons while their brains were chilled. Any memories that were retained through such a harrowing experience cannot depend on neural activity.

Amazingly, neurosurgeons sometimes chill the body and brain intentionally. In a dramatic medical procedure called Profound Hypothermia and Circulatory Arrest (PHCA), the heart is stopped and the entire body is cooled below 18 degrees Celsius, slowing life's processes to a glacial pace. PHCA is so risky that it's used only when surgery is required to correct a life-threatening condition. But the success rate is quite high, and patients usually survive with memories intact, even though their brains were effectively shut down during the procedure.

The success of PHCA supports a doctrine known as the "dual-trace" theory of memory. Persistent spiking is the trace of short-term memory, while persistent connections are the trace of long-term memory. To store information for long periods, the brain transfers it from activity to connections. To recall the information, the brain transfers it back from connections to activity.

The dual-trace theory explains why long-term memories can be retained without neural activity. Once activity induces Hebbian synaptic plasticity, the information is retained by the connections between the neurons in a cell assembly or synaptic chain. During recollection later on, the neurons are activated. But during the period between storage and recall, the activity pattern can be latent in the connections without actually being expressed.

It may seem inelegant to have two information stores. Wouldn't it be more effective for the brain to use just one? Computers, which are also used to store information, provide a helpful analogy. A computer contains two storage systems: the random access memory (RAM) and the hard drive. A document remains stored on your hard drive for long time periods. When you open the document in your word-processing program, your computer transfers the information from the hard drive to RAM. As you edit the document, the information in RAM is modified. When you save the document, your computer transfers the information from RAM back to the hard drive.

Since a computer was designed by human engineers, we know why it has two memory storage systems. The hard drive and the RAM both have their advantages. The hard drive has the virtue of stability; it can store information indefinitely, even if the power is turned off. In contrast, information in the RAM is volatile, easily lost. Imagine a power outage in the midst of editing, which causes all electrical signals inside the computer to cease. When you turn the computer on again ("reboot") and open the document, it will be intact — it was stored stably on the hard drive. But if you look closely, you will see that the document is the old version. Your edits, which were stored in the RAM, have disappeared.

If the hard drive is so stable, why use RAM at all? The answer is that RAM is speedy. Information in RAM can be modified more quickly than information on the hard drive. That's why it pays to transfer the document into RAM while editing and then transfer it back to the hard drive for safekeeping. It's often the case that the more stable something is, the more difficult it is to modify.

This tradeoff has been named the "stability–plasticity dilemma" by the theoretical neuroscientist Stephen Grossberg. Plato already recognized it in his dialogue *Theaetetus*. He explained memory failures as being caused by wax that is too hard or too soft. Some people have trouble storing memories, because their wax is too hard to be imprinted. Others have trouble retaining memories, because impressions are easily effaced from their too-soft wax. Only if wax is neither too hard nor too soft can it both take an impression and retain it.

The tradeoff between stability and plasticity may also explain why the brain uses two information stores. Like information in RAM, patterns of spiking change quickly and are suited to active manipulation of information during perception and thought. Because they are easily disturbed by new perceptions and thoughts, patterns of spiking are useful only for retaining information over short periods of time. Connections, in contrast, are analogous to the hard drive. Because connections change more slowly than spiking patterns, they are less suited to active manipulation of information. They are still plastic enough to store information, however, and stable enough to retain it for long durations. Hypothermia quenches neural activity, similar to the way that a power outage erases the RAM of a computer. Connections are left intact, so long-term memories survive. But recent information is lost, having not yet been transferred from activity to connections.

Can the stability–plasticity tradeoff also help us understand why the brain might use reconnection in addition to reweighting as a means for storing memories? Through Hebbian plasticity, neural spiking is continually altering synaptic strengths. Therefore the strength of a synapse is not so stable, and the memories stored by reweighting might not be either. This could explain why the memory of what you had for dinner yesterday will most probably fade. On the other hand, the existence of a synapse may be more stable than its strength. A memory stored by reweighting might be further stabilized by reconnection. This is likely the case for memories that endure for a lifetime, such as your name. Indelible memories may depend less on maintaining synaptic strengths at constant values and more on maintaining the existence of synapses. As a more stable but less plastic means of storing memories, reconnection may serve a complementary role to reweighting.

This chapter has been a mixture of empirical fact and theoretical speculation, biased uncomfortably toward the latter. We know for sure that reweighting and reconnection happen in the brain. Whether these phenomena create cell assemblies and synaptic chains is unclear, however. More generally, it has been difficult to prove that these phenomena are involved in any way in the storage of memories.

One promising method is to disable Hebbian synaptic plasticity in animals using drugs or genetic manipulations that interfere with the appropriate molecules at synapses, and then do behavioral experiments on the animals to see whether and how memory is impaired. Such experiments have already yielded fascinating and tantalizing evidence in support of connectionism. Unfortunately, the evidence is only indirect and suggestive. And its interpretation is complicated, because there is no perfect way of getting rid of Hebbian synaptic plasticity without creating other side effects.

The following parable is my attempt to illustrate the difficulties that neuroscientists face in testing theories of memory. Suppose that you are an alien from another planet. You find humans ugly and pathetic but are nevertheless curious about them. As part of your research you are spying on a particular man. He carries a notebook in his pocket. Every now and then he opens it and leaves marks on the pages with a pen. Sometimes he opens the notebook, looks at it briefly, and puts it back in his pocket.

You find this behavior puzzling, since you've never seen or heard of writing. Tens of millions of years ago your ancestors used writing, but that stage of evolution has been completely forgotten. After a great deal of thinking, you formulate the hypothesis that the man is using the notebook as a memory device.

One night, in order to test your hypothesis, you hide the book. In the morning the man spends a long time wandering about his house, looking under his bed, opening cabinets, and so on. For the rest of the day his behavior sometimes looks different, but only marginally so. You are feeling a bit discouraged, so you imagine other experiments to test your hypothesis: Cut just a few pages out of the book. Dunk it in water to erase the marks. Swap his notebook with someone else's.

The most direct test would be to read the writing in the notebook. By decoding the ink marks on the paper, you might be able to predict the events of the man's coming day. If your predictions turned out to be correct, that would be strong evidence that the notebook stores information. Unfortunately, you are now over twenty thousand years old and farsightedness has set in. Although your surveillance device allows you to look at the notebook, you can't see the writing clearly. (It's a bit far-fetched, but let's suppose that your alien civilization hasn't invented reading glasses or bifocals.)

Like you, the farsighted alien, neuroscientists want to test a hypothesis about memory. They believe that information is stored by modifying the connections between neurons. To test the hypothesis, they destroy the brain areas that contain the connections, just as you hide the notebook that contains the writing. They measure whether the brain area is activated when memory tasks are being done, just as you check whether the man pulls the notebook from his pocket when he needs to remember something.

Another strategy would be more direct and conclusive: attempt to read memories from connectomes. Look for the cell assembly and synaptic chain to see if they actually exist. Unfortunately, in the same way that your farsighted eyes can't even see the writing in the man's book clearly (much less decode it), neuroscientists can't see connectomes. That's why we need better technologies to understand the mysteries of memory.

Before I describe these emerging technologies and their potential applications, I need to talk about one more important factor that shapes connectomes. Experience may reweight and reconnect neurons, but genes shape connectomes as well. In fact, one of the most exciting prospects for connectomics is the promise of finally uncovering the interplay between the two. The connectome is where nature meets nurture.

PART III

NATURE AND NURTURE

SIX

The Forestry of the Genes

THE ANCIENT GREEKS compared human life to a slender thread — spun, measured, and cut by three goddesses called the Fates. Today biologists search for the secrets of human destiny in a different thread. The molecule known as DNA consists of two strands wound into a double helix. Each strand is a chain of smaller molecules called nucleotides, which come in four types designated by the letters A, C, G, and T. Your DNA spells out billions of these letters, in a sequence known as your genome. This sequence contains tens of thousands of shorter segments called genes.

It has been obvious throughout human history that children look a lot like their parents. When a baby is born, the comments start almost immediately — "She's got your eyes!" "He has your curly hair!" DNA provides an explanation. A child inherits half its genes from one parent and half from the other, and therefore inherits traits from both. Everyone accepts this idea for the body, but it's more controversial for the mind.

Perhaps the human mind is so malleable that it is shaped more by experiences than by genes, as Locke believed when he compared

the mind to white paper, ready to be inscribed. Then again, there's
no question that children often resemble their parents in more than
just looks. You can try to deny it when someone tells you, "The apple
doesn't fall far from the tree" or "You're a chip off the old block," but
there will come a day when you realize you just responded to a situation in *exactly* the way your father did three decades earlier. But of
course this anecdotal observation, while suggestive, won't prove anything. The similarity might be the result of upbringing rather than
genes.

These two explanations — genes and upbringing — were called "nature" and "nurture" by Francis Galton. Only in the twentieth century
did the nature–nurture debate finally move beyond philosophical as-
sertion and personal anecdote. Convincing evidence came from monozygotic (MZ) twins, who originated from a single zygote (fertilized egg
cell) and therefore share the same genome. Researchers identified and
studied MZ, or "identical," twins who were separated at an early age
and raised in different adoptive families. Their IQ scores turned out to
be as similar as their physical traits, such as height and weight. They
were much more similar than the IQ scores of two persons chosen
at random. The extra similarity can't be explained by shared environment, because these twins were raised in different adoptive families. It
can plausibly be explained by their shared genome. From this data, it
appears that genes influence IQ as strongly as they influence physical
traits.

This kind of comparison has been repeated for many other mental
traits beyond IQ. Personality tests are filled with questions like "I see
myself as someone who tends to find fault with others," to which the
test taker responds with an answer between 1 ("strongly disagree") and
5 ("strongly agree"). Twins score less similarly on personality tests than
on IQ tests, but their scores are still more similar than those of two
persons chosen at random, even if the twins were raised apart. This
means that personality is more malleable than IQ, but genetic factors
are still important.

For a long time, twin studies aroused intense opposition from believers in the power of nurture. By now, though, the studies have been

replicated so many times that there remains little room for argument. The psychologist Eric Turkheimer has promulgated the First Law of Behavior Genetics: "All human behavioral traits are heritable."

This law holds not only for mental differences between normal people but also for mental disorders. Early on, those trained in the psychoanalytic tradition believed that autistic children were the product of "refrigerator mothers." In a 1960 profile of Leo Kanner, the psychologist who first defined autism, *Time* magazine wrote: "All too often this child is the offspring of highly organized, professional parents, cold and rational — the type that Dr. Kanner describes as 'just happening to defrost enough to produce a child.'" But Kanner was actually ambivalent in his beliefs about the cause of autism. In the conclusion of the 1943 paper in which he originally defined autism, he noted that many of his patients had emotionally cold parents, but he went on to say that their condition was innate.

This leads us to another possible cause of autism: faulty genes. Researchers have explored this idea, too, by studying twins. If autism were completely determined by genetic factors, we'd expect MZ twins to both be autistic or both be normal. In fact, the agreement is not perfect. If one twin has autism, so does the other, with 60 to 90 percent probability. Since this concordance rate, as it is called, is less than 100 percent, autism is not *completely* determined by genes. Nevertheless, the rate is still high, and suggests that genetic factors are important for autism.

Of course, this statistic is not conclusive by itself. Because twins generally grow up in the same household, they tend to have similar experiences. If Kanner's "refrigerator mothers" were the cause of autism, that too would lead to high concordance rates. In the IQ studies, the effects of genes and environment were teased apart by studying MZ twins adopted and raised in separate households. It's difficult to locate such twins, and even more difficult to find such twins with autism, so geneticists have taken a different approach. They study twins raised together, and assess the importance of genes by comparing MZ twins with dizygotic (DZ), or "fraternal," twins. It turns out that the concordance rate for autism is relatively low in DZ twins, just 10 to 40 percent. This lower concordance rate is easily explained if autism is influenced by genetic factors, since DZ twins are genetically less similar than MZ twins. (DZ twins share 50 percent of their genes, while MZ twins share 100 percent.)

What about schizophrenia? The concordance rate is again lower for DZ twins (0 to 30 percent) than for MZ twins (40 to 65 percent). These numbers suggest that genetic factors are important for schizophrenia as well.

The studies of twins show that genes matter, but they do not explain why. Before I tackle the answer (or many answers) to this question, let me explain some things about genes.

You can think of a cell as an intricate machine built from molecular parts of many types. One of the main types is a class of molecules known as proteins. Some protein molecules can be structural elements, supporting the cell like the studs and joists of a wooden house frame. Other protein molecules perform functions on other molecules, much as workers in a factory handle parts. Many proteins combine both structural and functional roles. And the cell is more dynamic than most man-made machines, as many of its proteins move around from place to place.

It's commonly said that DNA is the blueprint of life, because it contains the instructions that cells follow to synthesize proteins. Just as DNA is a chain of nucleotides, a protein molecule is a chain of smaller molecules called amino acids, which come in twenty types. Each kind of protein is specified by a sequence of letters, but the alphabet contains twenty letters rather than the four used in DNA. This amino acid sequence is specified by a (mostly) contiguous string of letters—a gene—in your genome. To produce a protein molecule, the cell reads the nucleotide sequence of a gene and "translates" this into an amino acid sequence to synthesize a protein. (The dictionary for translation is known as the genetic code.) When a cell reads a gene and constructs a protein, it is said to "express" the gene.

You started your life as a single cell, an egg fertilized by a sperm.

This cell divided in two, and its progeny divided, and so on for many generations to produce the huge number of cells in your body. Every dividing cell replicated its DNA and passed on identical copies to its progeny. That's why every cell in your body contains the same genome. Why then do a liver cell and a heart cell look different and perform different functions? The answer is that cells of different types express different genes. Your genome contains tens of thousands of genes, each corresponding to a different kind of protein. Each type of cell expresses some of these genes but not others. Neurons are arguably the most complex type of cell in the body, so it's no surprise that many genes encode proteins that are exclusively or partially devoted to supporting functions in neurons. This is a preliminary answer to the question of why genes matter for the brain.

Your genome and mine are almost identical, conforming almost exactly to the sequence that was found by the Human Genome Project. But there are also slight differences, and the field of genomics is developing faster and cheaper technologies for detecting them. Sometimes the differences reside in single letters, while other times a longer stretch of letters is deleted or duplicated. If a genomic difference alters a gene, we can make a guess about the consequences if we know the function of the protein encoded by the gene.

By now you're familiar with the idea that mental function is based on spiking and secretion. Both processes involve many kinds of proteins. You've already encountered an important kind, the receptor molecules that sense neurotransmitter. These sit in the outer membrane of a neuron, partially protruding from the exterior of the cell. (Remember the kid floating in the inner tube?) Earlier I described the binding of a neurotransmitter molecule with a receptor as being like the insertion of a key into a lock. The metaphor goes even further for some receptors, which are a combination of a lock and a door. A small tunnel threads through the receptor molecule, connecting the inside of the neuron to the outside, but it's blocked by a doorlike structure most of the time. When the neurotransmitter binds to the receptor, the door opens for an instant, and electrical current can momentarily flow through the tunnel. In other words, the neurotransmitter acts like a key that opens a door, allowing electrical current to flow between the inside and outside of the neuron.

In general, we use the term *ion channel* for any type of protein containing a tunnel that passes electrical current through the membrane. (Ions are the electrically charged particles that conduct electricity in aqueous solutions.) Many types of ion channels are not receptors. Some of them enable the neuron to generate spikes; others have subtler effects on the electrical signals traversing neurons. If your genome contains an abnormal DNA sequence for a receptor or ion channel, it could be bad news for brain function. A disease caused by a defective DNA sequence for an ion channel is called a "channelopathy." Malfunctioning ion channels can lead to the uncontrolled spiking that we call epileptic seizures.

There are other types of proteins that package neurotransmitter into vesicles, as well as proteins that help release the contents of the vesicles into the synaptic cleft when triggered by a spike. Other proteins help degrade or recycle the neurotransmitter in the cleft, preventing it from lingering too long or drifting off to other synapses. This list is only the tip of the iceberg; it does not do justice to the vast array of proteins that serve spiking and secretion. Defects in any of these proteins could lead to brain disorders.

The possibilities for malfunction go way beyond that, however. On top of their present-day effects, defective genes might have made their mark in the past, when they caused the development of the young brain to go awry.

Roughly speaking, the brain grows and develops in four steps. Neurons are created, or "born," through the division of progenitor cells, migrate to their proper places in the brain, extend branches, and make connections. Disruption of any of these steps can lead to an abnormal brain.

What happens if the creation of neurons does not proceed successfully? In the city of Gujrat in Pakistan, there is a shrine to a seventeenth-century holy man named Shua Dulah. For centuries, babies born with abnormally small heads have been left at this shrine. In Pakistan they are known as *chuas,* which translates as "rat people," probably because their faces protrude in a somewhat ratlike way. The chuas are sometimes exploited by chua masters, who send them out to beg and then take the proceeds. The people tell various myths to explain the existence of chuas. One is the gruesome story that chuas are created by evil people who place clay or metal caps around the heads of babies, thereby retarding the growth of their brains.

In reality, the chuas are born with the disorder of congenital microcephaly. In the purest form, microcephaly vera, the only abnormality appears to be reduced brain size at birth. The cortex is smaller, but the pattern of folds and other architectural features are roughly normal. Not surprisingly, given the smaller cortex, microcephaly vera is accompanied by mental retardation.

Researchers have found that defects in a number of genes (with names like microcephalin or ASPM) can cause microcephaly vera. These genes encode proteins that control the birth of cortical neurons. Defects in them reduce the number of neurons and cause microcephaly. Because there are two copies of every gene, it's possible to carry one defective copy without showing any symptoms; the single correct copy is enough to make the brain grow normally. But when two carrier parents each pass on a defective copy to their child, he or she is born with microcephaly. This event would normally be rare, but in Pakistan it happens more frequently because of the high rate of intermarriage between cousins. (Since cousins are genetically related, it's more likely for them both to be carriers than it is for two people chosen at random.)

The second step of brain development, the migration of neurons to their proper places, can also be disrupted. In the disorder of lissencephaly (from the Greek roots for "smooth brain"), the cortex lacks the folds that normally give it a wrinkled appearance, and possesses other structural abnormalities visible in a microscope. The condition is usually accompanied by severe mental retardation and epilepsy. Lissencephalies are caused by mutations in genes that control neuronal migration during gestation.

These two steps in brain development occur in the prenatal brain. By the time a baby is born, the creation and migration of neurons are virtually complete. You may have heard that you were born with all the neurons that you will ever have. (There are only a few areas of the brain in which neurons still continue to be created after birth.) But this does not mean that brain development is over. Neurons continue to grow branches well after birth. This process is called the "wiring" of the brain, since axons and dendrites resemble wires. Axons have to grow the most, since they are much longer than dendrites. Imagine the tiny growing tip of an axon, known as a "growth cone" for its roughly conical shape. If a growth cone were blown up to human size, its travels would take it to the other side of a city. How is the growth cone able to navigate such long distances? Many neuroscientists study this phenomenon, and they've found that the growth cone acts like a dog sniffing its way home. The surfaces of neurons are coated with special guidance molecules that act like scents on the ground, and the interstitial spaces between neurons contain drifting guidance molecules that act like scents in the air. Growth cones are equipped with molecular sensors and can "smell" the guidance molecules to find their destination. The production of these molecules and their sensors is under genetic control. That's how genes guide the wiring of the brain.

If axons don't grow properly, "miswiring" results. Consider the corpus callosum, a thick bundle of 200 million axons connecting the left and right hemispheres of the cerebrum. In rare individuals, the callosum is either completely or partially missing. Fortunately, the impairments are much milder than in microcephaly. Such miswiring could be caused by defects in many genes, including those that control axon guidance.

For most of its journey through the brain, an axon grows straight, like the trunk of a tree. Once the growth cone reaches its final destination, the axon starts to branch. Scientists have reason to think that this final branching might not be so tightly controlled by genes. If this is the case, the detailed branching pattern of a neuron is largely random, although its overall shape might be genetically determined. Likewise, trees in a pine forest look similar because they come from the same genetic plan. No two trees match exactly branch for branch, however, because growth also involves randomness and is influenced by environmental conditions.

As the wires of the brain are laid down, neurons connect with each other by creating synapses. I hypothesized earlier that the process of synapse creation is random, happening with some probability whenever neurons contact each other. There is also room for genetic control, because neurons of different types might recognize each other through molecular cues and "decide" on that basis whether to connect. (I'll talk later about neuron types.)

So the initial connectome produced by very early development appears to be largely a product of genes and randomness. Scientists are still studying their relative contributions. According to one theory, genes exert their influence mostly by controlling how the brain wires up. Genes roughly determine the shape of a neuron, the region over which it extends branches. If there is an overlap between the regions spanned by two neurons, there is potential for connection between them. But whether they actually connect is not determined by genes. At first, it depends on random encounters of branches within the genetically defined regions, and on random creation of synapses at these encounters. But as development proceeds, experiences also start to shape the connectome. How exactly does this happen?

New synapses are created at a staggering rate in the infant brain. In Brodmann area 17 alone, over half a million per second are produced between two and four months of age. To accommodate the synapses, neurites increase in both number and length. Figure 25 illustrates the dramatic growth of dendritic branches from birth to two years of age.

I cautioned in Chapter 5 against thinking of adult learning as purely synapse creation. The same is true of the young brain, for development also *destroys* connections. When you were two years of age, you

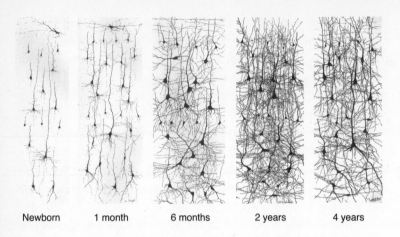

Newborn   1 month   6 months   2 years   4 years

FIGURE 25. *Dendrite growth from birth to age two, followed by pruning*

had far more synapses than you have now. By adulthood, the number
of synapses has dropped to 60 percent of its peak during the toddler
years. A similar rise and fall holds for the branches of neurons. Dendrites and axons grow exuberantly at first, but some branches are later
pruned away (compare the last two panels of Figure 25).

Why does the brain create so many synapses, only to destroy many
of them later? Actually, many so-called creative acts are misnamed,
because they involve both creation and destruction. When I'm writing
an article, I focus first on getting all my thoughts out onto the page,
even if the writing is embarrassingly bad. During this phase, the words
increase in number. After a rough draft is complete, further rewriting
or editing often shortens the piece. The final article ends up having
fewer words than the draft. As the saying goes, perfection is achieved
not when there is nothing left to add, but when there is nothing left to
take away.

Perhaps the early connectome is like a rough draft. I said above that
the initial wiring and the creation of connections are guided by genes
but also subject to randomness. And earlier I mentioned the theory
that synapse elimination in the adult brain is driven by weakening,
which in turn is driven by experience. By the same arguments, expe-

rience is likely to be the main driver of synapse elimination in the developing brain. And perhaps the elimination of many synapses from a branch leads to its pruning. These destructive processes refine the rough draft to produce the adult connectome.

This scenario is slightly misleading, however, because it suggests that creation and destruction occur in two distinct phases. The writing analogy clarifies why this is implausible. While working on a rough draft, I both add and remove words. There is *net* word creation because additions outnumber deletions. It's the other way around in the later phase of refinement, when the total number of words is decreasing. So it would be a mistake to think that before age two it's only synapse creation that occurs, and thereafter it's only synapse elimination. Net creation occurs early and net elimination occurs later, but both processes happen throughout life. Even in adulthood, when the total number of synapses remains roughly constant, both creation and elimination are taking place.

If synapse creation is mostly random while synapse elimination is driven by experience, shouldn't enriched cages cause synapse number to *decrease* in rats? Recall the finding of William Greenough and other researchers (mentioned in Chapter 5) — that synapses increase in number. We can only speculate, but here's one plausible scenario. Let's suppose that synapse elimination does happen at a greater rate in the brain of an enriched-cage rat, because it is learning more, but then, to replace the eliminated synapses, the brain steps up the creation of new ones. If creation more than compensates for elimination, the result is a net increase in synapse number. In this speculation, the increase in synapse number is the *effect* of learning rather than its cause.

The oxymoron *creative destruction* was central to the Austrian economist Joseph Schumpeter's theory of economic growth and progress. Its first word referred to the creation of new companies by entrepreneurs, and its second to the destruction of inefficient companies by bankruptcy. Brain development, writing an article, and economic growth all involve an intricate interplay between creation and destruction. Both processes are required for complex *patterns* of organization to evolve. When seen in this light, it verges on futile to measure progress by counting the total number of synapses in a brain, words in an article, or companies in an economy. It's the organization of the brain that matters, not the number of synapses.

By now you should have some appreciation for the intricacies of brain development. There are plenty of ways for such a complex process to go wrong. Disruption of the earliest steps of development, the creation and migration of neurons, is expected to cause abnormalities that are easy to see, such as microcephaly and lissencephaly. But disruption of the later steps of development could lead to *connectopathies,* disorders of neural connectivity. The total number of neurons and synapses would be normal, but they would be connected in a less than ideal way.

Remember the story of the Cray-1 supercomputer, which contained hundreds of thousands of wires totaling 67 miles in length? Remarkably, the first time it was powered up, it worked properly. The workers who built it had succeeded in connecting every single wire correctly. Your brain is far more complex, containing millions of miles of "wire." It's a wonder that any brain can ever develop correctly at all.

As I mentioned earlier, the corpus callosum fails to develop in rare individuals. This connectopathy is visible in an MRI scan because the callosum is ordinarily so large. But given our inability to see brain connectivity clearly, it's likely that the vast majority of connectopathies remain undiscovered. These will be revealed as our technologies for finding connectomes advance.

Earlier I zeroed in on the most puzzling aspect of autism and schizophrenia — the lack of a clear and consistent neuropathology. Studies of twins convinced researchers years ago that autism and schizophrenia have some basis in faulty genes. But exactly which of the tens of thousands of genes are faulty? Most researchers now suspect that many of the culprits are somehow involved in brain development. Autism and schizophrenia are said to be *neurodevelopmental* disorders, in which the brain fails to grow normally. They are fundamentally different from neurodegenerative disorders like Alzheimer's disease, in which an originally normal brain starts to fall apart.

What is the evidence behind this suspicion? The case is more clearcut for autism, as its symptoms are detected in early childhood. Whatever the neuropathology may be, it must have emerged during gestation and infancy, when the brain was growing most rapidly. Earlier I mentioned that autistic children have larger brains on average. Looking at brain growth over time reveals a more complex picture. The autistic brain is slightly smaller than average at birth, larger than average from age two to age five, and average again by adulthood. In other words, the *rate* of brain growth is abnormal in autistic children. This suggests a developmental abnormality, but conclusive proof would require identifying a clear and consistent neuropathology that emerges in the womb or during infancy.

In the first half of the twentieth century, researchers did not believe that schizophrenia was neurodevelopmental. They hypothesized that the schizophrenic brain was normal during childhood and that it started to degenerate in late adolescence or young adulthood, triggering the first episode of psychosis. But they failed to find neuropathologies that should accompany a degenerating brain, so the hypothesis had to be abandoned.

Today many researchers speculate that schizophrenia, like autism, is a neurodevelopmental disorder. It turns out that many schizophrenics experienced slight delays in learning to talk, move, and socialize, so perhaps their brains were already slightly abnormal in childhood. Their brain development might even have veered off course in the womb: Statistical studies suggest that pregnant mothers exposed to famine or viral infection are more likely to give birth to children who later develop schizophrenia.

So here's what researchers believe: Autism and schizophrenia are caused by some neuropathology, which is caused by abnormal brain development, which is caused by some combination of abnormal genetic and environmental influences. Neuroscientists are just beginning to find the genes, which could help them close in on the relevant

developmental processes. This sounds encouraging, but I'm embarrassed to admit that the most important question has still not been
answered: What is the neuropathology? Without data, theories have
abounded. Since these are far too numerous to review exhaustively, I'll
focus on the one that makes the most sense to me — the theory that
autism and schizophrenia are connectopathies.

Recall that the autistic brain grows faster than normal in early childhood. The overgrowth is somewhat greater in the frontal cortex than
in other lobes, perhaps because too many connections are created between neurons there. In addition, researchers speculate that too *few*
connections are created between the frontal cortex and other regions
of the brain.

It's distressing to realize that this theory of autism is based on phrenological evidence and couched in phrenological terms. As I've mentioned, the enlargement of the autistic brain is only statistical, governing only averages. Diagnosing autism in an individual child based on
the size of the brain or its regions would be grossly inaccurate. Statements about "too many" or "too few" connections are just as crudely
phrenological as "too large" or "too small." If autism is caused by a connectopathy, the difference will probably be found in the organization
of connections, rather than in their overall number. The connectopathy would be invisible to our current technologies; hence the failure to
find a clear neuropathology for autism.

Could schizophrenia, too, be caused by a connectopathy? Here the
most tantalizing evidence comes from studies of synapse elimination. Earlier I mentioned that adults have fewer synapses than babies,
but I did not describe exactly when the reduction occurs. Researchers have found that synapse number declines rapidly after the peak in
infancy, stays roughly constant during childhood, and drops rapidly
again in adolescence. Perhaps something goes wrong in the schizophrenic brain during this second reduction. The defect is probably not
as simple as too few or too many synapses, as that kind of neuropathology would have been detected by now. Maybe the wrong synapses
are eliminated, and this pushes the brain over the edge to psychosis.

Finding a clear and consistent neuropathology should be a central goal of research on autism and schizophrenia. We will need to go beyond phrenological methods if these disorders are connectopathies; we will need the technologies of connectomics. In fact, I believe that studying autism and schizophrenia without connectomics is like studying infectious diseases without the microscope. Seeing the microbes that cause disease is not by itself a cure, but it accelerates research toward one. Similarly, finding a neuropathology that is truly distinctive of a mental disorder is not a cure by itself, but it's a step in the right direction.

For the sake of argument, however, let's consider the opposing view. Maybe searching for neuropathology is a waste of time. A genomics enthusiast might say that autism is caused by defective genes, so we should focus on finding them and not waste time with connectomes.

Indeed, the rapid progress of genomics is stunning. When genomic technologies were slow and expensive, researchers focused on a few rare families with a history of many afflicted members. Now it's possible to rapidly screen the genomes of large populations to find abnormalities. Researchers have discovered abnormalities in many different genes associated with autism and schizophrenia. This is exciting progress, but there are also limitations.

Genomics can predict with high confidence that a child who is born with certain genetic defects will develop autism or schizophrenia. But it cannot predict the vast majority of cases, because no single known defect can account for more than 1 or 2 percent of all cases, and most account for far fewer. In this sense, genomics is currently ineffective for predicting autism or schizophrenia in individuals, much as neophrenology cannot predict the IQ of individuals.

Genetic testing is much more successful at predicting Huntington's disease (HD), a neurodegenerative disorder that typically strikes in middle age. HD begins with random involuntary, jerky movements and eventually progresses to cognitive decline and dementia. Because only one gene is involved, HD is much simpler to predict than autism. An abnormal version of the gene can be detected by a highly accurate

DNA test. A positive result means that the individual will develop HD, and a negative result means that the individual will not.

Understanding the genetics of autism and schizophrenia is much trickier, given that so many genes are involved. One way forward is to say that autism is actually composed of a large number of autisms, each one caused by a different defective gene. We could study each autism independently and develop a different treatment for each one. This strategy is being pursued by many researchers now, and I expect it will be the most successful one in the short run. But in the long run a complementary strategy will also be fruitful. It may be the case that diverse genetic defects all produce the same neuropathology. I believe we should focus on identifying that neuropathology and treating it.

A genomics enthusiast might argue that treating the neuropathology is not the right approach, because it doesn't strike at the cause. If defective genes cause mental disorders, we should use gene therapy to replace the bad copy of the gene with a good copy. Researchers have experimented with this strategy by engineering animals with genetic defects that lead to brain disorders. In some cases, they have had remarkable success in treating adult animals by correcting the genetic defect. Such research could eventually lead to therapies for human patients. But this strategy may not always work, or may be only partially successful. If the genetic defect primarily disrupts brain function in the present, then correcting it should solve the problem. But if the defect did most of its damage in the past by altering brain development, correcting it now may not be as helpful.

An analogy may clarify the issue. Imagine that you're suffering from depression because your marriage is breaking up. You go to an old-fashioned psychoanalyst for help, and you're told that your problems spring from the bad relationship you had with your mother when you were growing up. That may be true, but does this insight really help you fix the problem? Now that you're all grown up, replacing your mother with an adoptive mother would have little effect.

Saying that mental disorders are caused by defective genes is the modern way of blaming one's parents: It's not obvious how to use this historical explanation as a basis for treatment. Gene therapy on an adult with a brain that failed to develop normally might be as ineffective as replacing an adult's mother.

Now suppose that a mental disorder is caused by a connectopathy. A true cure requires correcting the abnormal connectivity. So the obvious question is: How much can we change our connectomes, and what is the best way to do it?

SEVEN

Renewing Our Potential

IN THE GAME of life, you are dealt genes. You can't change your genome; it's the hand you must play. The genomic worldview is pessimistic, constrained on all sides. In contrast, your connectome changes throughout life, and you have some control over that process. The connectome bears an optimistic message of possibility and potential. Or does it? How much can we really change ourselves?

The Serenity Prayer, quoted at the beginning of Chapter 2, echoes the sentiments of an older rhyme:

> *For every ailment under the sun*
> *There is a remedy, or there is none;*
> *If there be one, try to find it;*
> *If there be none, never mind it.*

That kind of mixed message is also on display in the self-help section of your local bookstore. Browse for a few minutes and you'll come across many books that don't tell you how to change; instead, they teach resignation. If you're persuaded that you can't possibly change your spouse, you may stop nagging and learn to be happy with your marriage. If you believe that your weight is genetically determined, you may cease dieting and enjoy eating once again. On the other end of the spectrum, diet books like *I Can Make You Thin* and *Master Your Metabolism* are titled to inspire optimism about losing weight. In his guide to self-help books, *What You Can Change and What You Can't*, the psychologist Martin Seligman lays out the empirical evidence for pessimism. Only 5 or 10 percent of people actually achieve long-term weight loss by dieting. That's a depressingly low number.

So is change really possible? The twin studies showed that genes may influence human behavior but do not completely determine it. Nevertheless, another type of determinism has emerged, this one based on the brain, and almost as pessimistic. "Johnny's just that way — he's wired differently," you hear people say. Such *connectome determinism* denies the possibility of significant personal change after childhood. The idea is that connectomes may start out malleable but become fixed by adulthood, in line with the old Jesuit saying, "Give me the child until he is seven and I'll give you the man."

The most obvious implication of connectome determinism is that changing people should be easiest in the first years of life. The construction of a brain is a long and complex process. Surely it's more effective to intervene during the early stages of construction, rather than later on. While a house is being built, it's relatively easy to deviate from the architect's original blueprint. But as anyone who has remodeled a house knows, it's much harder to make major changes after the house is finished. If you've tried to learn a foreign language as an adult, you may have found it a struggle. Even if you were successful, you probably didn't end up sounding like a native speaker. Since children seem to learn second languages effortlessly, their brains appear to be more malleable. But does this idea really generalize to mental abilities other than language?

In 1997, then–First Lady Hillary Clinton hosted a conference at the White House entitled "What New Research on the Brain Tells Us about Our Youngest Children." Enthusiasts of the "zero-to-three movement" gathered to hear claims that neuroscience had proven the effectiveness of intervening during the first three years of life. At the conference was the actor and director Rob Reiner, who started the I Am Your Child Foundation, also in 1997. He was beginning to create a series of educational videos for parents about the principles of child-rearing. The inaugural title was "The First Years Last Forever," which sounded ominously deterministic.

Actually, neuroscience has been unable to confirm or deny such claims, because it's been difficult to identify exactly what changes in the brain cause learning. Could the zero-to-three movement base its claims of determinism on the neo-phrenological theory that learning is caused by synapse creation? (Let's ignore the considerable evidence against this theory, for the sake of argument.) The answer would be yes if synapse creation were impossible in adults. But William Greenough and other researchers showed that connection number still increases even when adult rats are placed in enriched cages. The rate was slower than in young rats, but still substantial. And remember the MRI studies of the cortex in people learning to juggle? Thickening occurred in the elderly as well as young adults. Finally, watching synapses through a microscope has shown that reconnection still continues in the brains of adult rats, as mentioned previously. Neuroscientists have not demonstrated a drop in reconnection with age as dramatic as the decrease in language-learning ability. Therefore, the first form of connectome determinism, "reconnection denial," does not seem tenable.

A second form has emerged, however: "rewiring denial." The "wires" of the brain are laid down in early life, as neurons extend axons and dendrites. Retraction of branches also occurs during development. Using microscopy, researchers have been able to capture videos of these remarkable processes. Often the tip of an axon makes a synapse onto a dendrite, gripping as if the synapse were like a hand. The creation of such a synapse appears to stimulate the axon to grow further, but if such a synapse is eliminated, the axon loses its hold and retracts. In general, it seems that axonal branches can't be stable unless they make synapses. Although growth and retraction are highly dynamic in the young brain, rewiring deniers believe that they grind to a halt in the adult brain. The wires can be reconnected in new ways by synapses, and synapses can be reweighted by changing their strengths, but the wires themselves are fixed.

Rewiring is hotly debated because of its suspected role in remapping, the dramatic changes in function observed after brain injury or amputation. To understand the importance of rewiring, we need to revisit a more fundamental question: What defines the function of a brain region?

The whole notion of a brain region with a well-defined function implicitly depends on an empirical fact. Through measurements of neural spiking, it has been shown that neurons near each other in the brain (neighboring cell bodies) tend to have similar functions. One can imagine a different kind of brain in which neurons are chaotically scattered without any regard for their functions. It wouldn't make sense to divide such a brain into regions.

But why do the neurons in a region have similar functions? One reason is that most connections in the brain are between nearby neurons. This means neurons in a region "listen" mainly to each other, so we'd expect them to have similar functions, much as we'd expect less diversity of opinions among a group of people who mainly keep to themselves. This is part of the story, but not all of it.

The brain also contains some connections between distant neurons. In effect, neurons in the same region "listen" to neurons in other regions as well as each other. Couldn't these faraway sources of input lead to diversity? Indeed they could if they were distributed all over the brain, but in fact they are typically confined to a limited number of regions. Returning to the social analogy, you could imagine a brain region as a group of people who listen to the outside world a bit, but only by reading the same newspapers and watching the same television shows. These external influences are so narrow that they don't lead to diversity either.

Why are long-range connections constrained in this way? The answer has to do with the organization of brain wiring. Most pairs of regions lack axons running between them, so their neurons have no way of connecting with each other. In other words, any given region is wired to a limited set of source and target regions. This set has been called a "connectional fingerprint," as it appears to be unique for each region. The fingerprint is often highly informative about the region's function. For example, the reason that Brodmann area 3 mediates bodily sensations, a function I mentioned earlier, is that this area is wired to pathways bringing touch, temperature, and pain signals from the spinal cord. Similarly, the reason that Brodmann area 4 controls movements of the body is that this area sends many axons to the spinal cord, which in turn is wired to the muscles of the body.

These examples suggest that a region's function depends greatly on its wiring with other regions. If that's true, altering the wiring could change the function. Remarkably, this principle has been demonstrated by "rewiring" a nominally auditory area of the cortex to serve the function of vision. The first step was taken in 1973 by Gerald Schneider, who discovered an ingenious method to reroute axons growing in the brains of newborn hamsters. By damaging certain brain regions, he diverted retinal axons from their normal target in a visual pathway to an alternative destination in an auditory pathway. This had the effect of sending visual signals to a cortical area that is normally auditory.

The functional consequences of this rewiring were investigated in the 1990s by Mriganka Sur and his collaborators. After repeating Schneider's procedure in ferrets, they showed that neurons in the auditory cortex now responded to visual stimulation. Furthermore, the ferrets could still see even after the visual cortex was disabled, presumably by using their auditory cortex. Both pieces of evidence implied that the auditory cortex had changed its function to be visual. Similar "cross-modal" plasticity has also been observed in humans. For example, in those who are blind from an early age, the visual cortex is activated when they read Braille with their fingertips.

Such findings are consistent with Lashley's doctrine of equipotentiality, but they suggest an important qualification: A cortical area indeed has the potential to learn any function, but only if the necessary wiring with other brain regions exists. If every area in the cortex were wired to every other area (and to all other regions outside the cortex), then equipotentiality might hold without any provisos. Wouldn't the brain be far more versatile and resilient if its wiring were "all to all"? Maybe so, but it would also swell to gigantic proportions. All those wires take up space, as well as consume energy. The brain has evidently evolved to economize, which is why the wiring between regions is selective.

The Schneider and Sur experiments induced young brains to wire up differently. What about the adult brain? If the wiring between regions becomes fixed in adulthood, that would constrain the potential for change. Conversely, if the adult brain could rewire, it would have more potential to recover from injury or disease. This is why researchers so badly want to know whether rewiring is possible in adulthood, and also find therapies to promote the phenomenon.

In 1970, a thirteen-year-old girl came to the attention of social workers in Los Angeles. She was mute, disturbed, and severely underdeveloped. Genie (a pseudonym) had been a victim of terrible abuse. She had spent her entire life in isolation, tied up or otherwise confined to a single room by her father. Her case aroused great public attention and sympathy. Doctors and researchers hoped that she could recover from her traumatic childhood, and they resolved to help her learn language and other social behaviors.

Coincidentally, 1970 also saw the premiere of François Truffaut's film *L'Enfant Sauvage*, about the Wild Boy of Aveyron. Victor was discovered around 1800 wandering naked and alone in the woods of France. Efforts were made to "civilize" him, but he never learned to speak more than a few words. History has recorded other examples of so-called feral children, who grew up lacking exposure to human love and affection. No feral child was ever able to learn language.

Cases like Victor's suggested the existence of a *critical period* for the learning of language and social behaviors. Deprived of the opportunity to learn during the critical period, feral children could not learn these behaviors later on. In metaphorical terms, the door to learning hangs open during the critical period; then it swings shut and locks. While this interpretation is plausible, too little is known about feral children for it to be scientifically rigorous.

When Genie was found, researchers hoped that her case might overturn the theory of the critical period. They resolved to study Genie and rehabilitate her at the same time. She made some encouraging progress in learning language, but eventually funding for the research dried up. Then Genie's life took a tragic turn as she passed through a series of foster homes and seemed to regress.

Around the time the research ended, scientific papers reported that Genie was still learning new words but was struggling with syntax. According to later popular accounts, the researchers became discouraged, predicting that she would never learn real sentence structure. Whether Genie would have progressed further will never be known. She provided some evidence for a critical period in language learning, but it is difficult to draw firm scientific conclusions, however heartbreaking and gripping her case may be.

Optometrists encounter less harrowing forms of deprivation all the time. Weak vision in one eye often goes unnoticed if the other eye provides clear sight. Wearing eyeglasses or having a cataract removed easily corrects the problem with the eye. Nevertheless, the patient may still not see clearly with the corrected eye, or be stereo-blind, because there is still something wrong with the brain. (At a movie theater you've probably tried 3D glasses, which give a sensation of depth by presenting slightly different images to the two eyes. Those who can't perceive 3D in this way are said to be stereo-blind.) The condition, known as amblyopia to specialists, is nicknamed "lazy eye," but the disorder involves the brain as well as the eye.

Amblyopia suggests that we are not simply born with the ability to see; we must also learn from experience, and there is a critical period for this process. If the brain is deprived of normal visual stimulation from one eye during this limited time window, it does not develop normally. The effect is irreversible in adulthood. Children, however, recover normal vision if amblyopia is detected and treated early; their brains are still malleable. On the flip side, if an adult develops poor vision in a single eye, it has no lasting effect on the brain. Correcting the eye produces full recovery.

Amblyopia seems to document the claim made in the title of Rob Reiner's video, *The First Years Last Forever*. Early intervention is crucial, as the zero-to-three movement contends. Amblyopia treatments suggest that the brain becomes less malleable after the critical period. But can that be shown directly by neuroscience? How exactly do poor vision and corrected vision change the brain during the critical period, and why don't these changes happen later on?

In the 1960s and 1970s David Hubel and Torsten Wiesel investigated these questions with experiments on kittens. To simulate amblyopia, they occluded vision in one eye, a condition they called "monocular deprivation." Several months later they removed the occlusion and tested visual capability. The kittens could not see well with the previously deprived eye, much like human patients with amblyopia. To find out what had changed in the brain, Hubel and Wiesel recorded spikes from neurons in Brodmann area 17. Since this cortical area is important for vision, it's also known as primary visual cortex or V1. They measured the responsiveness of each neuron to visual stimulation of the left eye alone, and of the right eye alone. Few neurons responded to stimulation of the previously deprived eye.

The functions of V1 neurons had been altered by monocular deprivation. Could this have been caused by a connectome change? That's a good guess if we believe the connectionist mantra that the function of a neuron is chiefly defined by its connections with other neurons. In the 1990s Antonella Antonini and Michael Stryker provided evidence pointing to the rewiring of axons bringing visual information into V1. Each incoming axon is monocular, meaning that it carries signals from just one eye. Depriving one eye caused its axons to retract dramatically, and the other eye's axons to grow. In effect, rewiring eliminated pathways from the deprived eye to V1, and created new pathways from the other eye to V1. This plausibly explained why Hubel and Wiesel had observed few V1 neurons responsive to the previously deprived eye.

Rewiring of V1 was important because it identified a connectome change that could be the cause of learning. Since rewiring both created and eliminated synapses and pathways, it served as another counterexample to the neo-phrenological idea that learning is simply the creation of synapses.

Antonini and Stryker were also able to address another question: Why is the brain less malleable after the critical period? Hubel and Wiesel had shown that monocular deprivation induced V1 changes in young kittens but not in adults. Once induced, the changes were reversible while the kittens were young, but became irreversible in adulthood. Antonini and Stryker explained this by showing that monocular deprivation in adulthood did not rewire V1. Furthermore, rewiring induced during the critical period was reversible if monocular deprivation was ended early, but not if it was ended late.

Antonini and Stryker's research would seem to support the case for early intervention, as recommended by the zero-to-three movement. But an important pitfall of this argument has been pointed out by William Greenough, who discovered the increases in neural connections produced by environmental enrichment in rat brains. Amblyopia, like Genie's lonely upbringing, *deprived* children of normal experiences. It suggested the existence of a critical period for deprivation. Does it necessarily follow that there is also a critical period for enriching childhood with special experiences?

Greenough and his colleagues say it doesn't. Since experiences like visual stimulation and exposure to language were normally available to all children throughout human history, brain development "expects" to encounter them, and has evolved to rely heavily upon them. On the other hand, experiences like reading books were not available to our ancient ancestors. Brain development could not have evolved to depend upon them. That's why adults can still learn to read, even if they did not have the opportunity in childhood.

What the zero-to-three movement really needs is an example of a critical period for learning from *altered* experience — an example that goes beyond mere deprivation. One such experiment was pioneered in 1897 by the American psychologist George Stratton. He fastened a homemade telescope to his face and placed opaque materials around the eyepiece so that no other light could enter his eyes. The telescope was designed not to magnify images but to invert them. It turned the world upside down and also reversed left and right like a mirror. Stratton heroically wore the telescope twelve hours a day and blindfolded himself when he took it off.

As you can well imagine, Stratton was extremely disoriented at first, even nauseated. His vision conflicted with his movements. If he tried to reach for an object to his side, he would use the wrong hand. When he corrected himself and used the other hand, even the simple act of pouring milk into a glass was exhausting. His vision also conflicted with his hearing: "As I sat in the garden, a friend who was talking with me began to throw some pebbles into the distance to one side of my actual view. The sound of the stones striking the ground came, oddly enough, from the opposite direction from that in which I had seen them pass out of my sight, and from which I involuntarily expected to catch the sound." But by the time Stratton ended the experiment, on the eighth day, he was moving with greater ease, and his vision and hearing had harmonized: "The fire, for instance, sputtered where I saw it. The tapping of my pencil on the arm of my chair seemed without question to issue from the visible pencil."

What Stratton had discovered is that the brain can recalibrate vision, hearing, and movement to resolve conflicts between them. Eye surgeons have encountered a similar recalibration in patients with strabismus. This condition, more commonly known as "crossed eyes," is sometimes corrected by surgery on eye muscles to rotate the eye. Turning the eye in this manner changes the vision of the patients, effectively rotating the world around them. The rotation is revealed by a simple experiment, in which patients are requested to point in the direction of a visual target while not being allowed to see their pointing arm. They consistently point to one side of the target, because their movements now conflict with their changed vision. But if they are tested again a few days after surgery, the pointing errors are reduced, showing that the brain is recalibrating.

What happens in the brains of patients as they adapt to strabismus surgery? Starting in the 1980s, Eric Knudsen and his collaborators addressed this question with experiments on barn owls. They used special eyeglasses that rotated the world 23 degrees to the right by bending light rays to one side. This mimicked the rotation of the visual world produced by strabismus surgery. (In fact, similar eyeglasses are sometimes used as a treatment for severe strabismus.) Owls raised with these eyeglasses behaved in a way that looked skewed to an observer. If they heard a sound, they turned their heads to the right of the source. This skewed behavior enabled them to look at the source, as it compensated for the rotation caused by the eyeglasses.

To study the neural basis of this behavioral change, Knudsen and his collaborators examined the inferior colliculus. This part of the brain is important for computing the direction of a sound based on comparing signals from the left and right ears. Much as there is a map of the body in Brodmann areas 3 and 4 (the sensory and motor "homunculi"), there is a map of the external world in the inferior colliculus. By recording spikes from neurons in this structure, Knudsen and his collaborators showed that the inferior colliculus map was displaced in a direction consistent with the skewed-looking behavior. They also showed that incoming axons shifted over in the map, suggesting that remapping had been caused by rewiring.

Knudsen and his collaborators further demonstrated a critical period for learning by applying and removing the eyeglasses at different ages. Placing the eyeglasses on adult owls raised normally did not produce a change in looking behavior. If young owls were raised with eyeglasses, the effects were reversible if the eyeglasses were removed early but not if removed in adulthood.

Based on the examples of the inferior colliculus and V1, it seems we can deny the possibility of rewiring in the adult brain. This might explain why adults have greater difficulty adapting to change. I mentioned in Chapter 2 that adults do not recover from hemispherectomy as well as children do. More generally, the Kennard Principle states that the earlier the brain damage, the greater the recovery of function. This principle has been criticized as simplistic, since exceptions are well-known, but it has some element of truth. It follows from rewiring denial, because rewiring is an important mechanism for remapping.

At the same time, the doctrine of rewiring denial is still under attack. Researchers using microscopes to monitor axons over long time periods in living brains have shown that new branches can grow in adults. The experiments are controversial, but there is a growing consensus that at least short growths are possible, though long extensions might not be. Some suspect that such rewiring is responsible for the cortical remapping that accompanies phantom limbs, although there is still little conclusive evidence.

Other researchers are challenging the concept of the critical period, saying that the effects of early deprivation may be more reversible than was previously thought. The conventional wisdom has been that it's impossible to acquire stereo vision in adulthood. In her book *Fixing My Gaze*, the neuroscientist Susan Barry relates how she acquired some stereo vision in her forties, after a lifetime of stereo blindness caused by childhood strabismus. She was able to do this by subjecting herself to a special regimen that trained her vision.

Barry's success suggests that the effects of critical-period experience are only difficult, not impossible, to reverse. Antonini and Stryker seemed to demonstrate convincingly that V1 lost its potential for change in adulthood, because rewiring ceased. This seemingly open-and-shut case has recently been challenged by the discovery of several treatments that restore plasticity to adult V1. Researchers have employed four weeks of the antidepressant medication fluoxetine (better known by the trade name Prozac), pretreatment with ten days of darkness, or simple environmental enrichment in the style of Rosenzweig. These treatments appear to extend the critical period into adulthood or eliminate it altogether.

Knudsen and his collaborators initially emphasized the failure of adult owls to adjust to rotation of the visual world. But later experiments sent a more optimistic message. The owls wore a sequence of eyeglasses, each of which rotated the world by a progressively larger angle. Over time, the owls eventually adapted to the same 23-degree rotation that the young owls could handle in one giant adjustment. The finding supported the general idea that adults can learn as much as juveniles, if training is done correctly.

Optimism about adult brain plasticity is currently in vogue. In the 1990s the zero-to-three movement contrasted the rigidity of the adult brain with the flexibility of the infant brain. Now the pendulum has swung to the other extreme. In his book *The Brain That Changes Itself: Stories of Personal Triumph from the Frontiers of Brain Science*, Norman Doidge tells inspiring stories about adults who have managed amazing recoveries from neurological problems. He argues that the brain is exceedingly plastic, much more than neuroscientists and physicians ever thought.

Of course the truth lies somewhere in between. It's incorrect to flatly deny the possibility of adult rewiring, but such denials might hold water if they were qualified with conditions. For example, they could be restricted to specific types of branches growing from certain neurons toward others, or from certain regions to others. And it's simplistic to regard rewiring as a single phenomenon. Rewiring actually encompasses a large number of processes involved in the growth and retraction of neurites. A more refined denial of rewiring might focus on just one of the processes included in this catchall term.

Since denials are conditional rather than absolute, they might be sidestepped by the right kind of training program, as Knudsen showed. And it appears that brain injury facilitates rewiring by releasing axonal growth mechanisms that are normally suppressed by certain molecules. Future drug therapies might target these molecules, enabling the brain to rewire in ways that are not currently possible.

Because of our crude experimental techniques, only drastic kinds of rewiring have been detectable. That's why neuroscientists resorted to the rather extreme experiences of monocular deprivation and Stratton-type eyeglasses. The still-invisible, subtler kinds of rewiring could well be important for more normal types of learning. Simply by providing a clearer picture of the phenomenon, connectomics is bound to aid research in this area.

In 1999 a bitter fight erupted between two neuroscientists. In one corner stood the defending champion, Pasko Rakic of Yale University. Starting in the 1970s, his famous papers had firmly established a dogma: No new neurons are added to the mammalian brain after birth, or at least after puberty. The upstart was Elizabeth Gould of Princeton University, who had astounded her colleagues by reporting new neurons in the neocortex of adult monkeys. (Most of the cerebral cortex consists of neocortex, the part mapped by Brodmann.) Her discovery was hailed by the *New York Times* as the "most startling" of the decade.

It's not hard to understand why this face-off between two professors ended up on the front page. It's amazing when the body repairs itself. Skin wounds heal, leaving only a scar. Of all the internal organs, the liver is the champion at self-repair, growing back even if two-thirds of it has been removed. If the adult neocortex could add new neurons, that would mean the brain has more capacity to heal itself than anyone expected.

In the end, neither contender could be declared the undisputed champion. The "no new neurons" dogma prevailed in the neocortex. However, Rakic himself was forced to concede that neurons are continually added to two regions of the adult brain, the hippocampus and the olfactory bulb. (The olfactory bulb is for the nose what the retina is for the eye, and the hippocampus is a major non-neocortical part of the cortex.)

Since new neurons normally appear in these two regions, even in the absence of injury, they presumably aren't for healing. Perhaps they enhance learning potential, much as new synapses were hypothesized to increase memory capacity by adding potential to learn new associations. The hippocampus belongs to the medial temporal lobe, in which the Jennifer Aniston neuron was found. Some researchers believe that the hippocampus serves as the "gateway" to memory; they theorize that it stores information first and later transfers it to other regions like the neocortex. If this is the case, the hippocampus might need to be extremely plastic, and new neurons would endow it with extra plasticity. Similarly, the olfactory bulb might use new neurons to help store memories of smells.

According to neural Darwinism, synapse elimination works in tandem with creation to store memories. Likewise, we'd expect the creation of neurons to be accompanied by a parallel process of elimination. This pattern holds true for many types of cells, which die throughout the body during development. Such death is said to be "programmed," because it resembles suicide. Cells naturally contain self-destruct mechanisms and can initiate them when triggered by the appropriate stimuli.

You might think that your hand grew fingers by adding cells. No — actually, cell death etched away at your embryonic hand to create spaces between your fingers. If this process fails to happen properly, a baby is born with fingers fused together, a minor birth defect that can be corrected by surgery. So cell death acts like a sculptor, chiseling away material rather than adding it.

This is the case for the brain as well as the body. Roughly as many of your neurons died as survived while you floated in the womb. It may seem wasteful to create so many neurons and then kill them off. But if "survival of the fittest" was an effective way of dealing with synapses, it might also work well for neurons. Perhaps the developing nervous system refines itself through survival of neurons that make the "right" connections, coupled with elimination of those that don't. This Darwinian interpretation has been proposed not only for development but also for creation and elimination of neurons in adulthood, which I'll call *regeneration*.

If regeneration is so great for learning, why doesn't the neocortex do it? Perhaps this structure needs more stability to retain what has already been learned, and must settle for less plasticity in order to achieve that. But Gould's report of new neocortical neurons is not alone in the literature; similar studies have been published sporadically since the 1960s. Perhaps these scattered papers contain some grain of truth that's contrary to the current thinking among neuroscientists.

We could resolve the controversy by hypothesizing that the degree of neocortical plasticity depends on the nature of the animal's environment. Plasticity might well plummet in captivity, for confinement in small cages must be dull compared with life in the wild, and presumably demands little learning. The brain could respond by minimizing the creation of neurons, and most of those created might not survive elimination for long. In this scenario, new neurons indeed exist, but in small and fluctuating numbers that are hard to see, which would explain why researchers are split. It's entirely possible that more natural living conditions would foster learning and plasticity, and new neurons would become more numerous.

You might not be convinced by this speculation, but it illustrates a general moral of the Rakic–Gould story: We should be cautious about blanket denials of regeneration, rewiring, or other types of connectome change. A denial has to be accompanied by qualifications if it's to be taken seriously. Furthermore, the denial may well cease to be valid under some other conditions.

As neuroscientists have learned more about regeneration, simply counting the number of new neurons has become too crude. We'd like to know why certain neurons survive while others are eliminated. In the Darwinian theory, the survivors are the ones that manage to integrate into the network of old neurons by making the right connections. But we have little idea what "right" means, and there is little prospect of finding out unless we can see connections. That's why connectomics will be important for figuring out whether and how regeneration serves learning.

I've talked about four types of connectome change — reweighting, reconnection, rewiring, and regeneration. The four R's play a large role in improving "normal" brains and healing diseased or injured ones. Realizing the full potential of the four R's is arguably the most important goal of neuroscience. Denials of one or more of them were the basis of past claims of connectome determinism. We now know that such claims are too simplistic to be true, unless they come with qualifications.

Furthermore, the potential of the four R's is not fixed. Earlier I mentioned that the brain can increase axonal growth after injury. In addition, damage to the neocortex is known to attract newly born neurons, which migrate into the zone of injury and become another exception to the "no new neurons" rule. These effects of injury are mediated by molecules that are currently being researched. In principle we should be able to promote the four R's through artificial means, by manipulating such molecules. That's the way genes exert their influence on connectomes, and future drugs will do the same. But the four R's are also guided by experiences, so finer control will be achieved by supplementing molecular manipulations with training regimens.

This agenda for a neuroscience of change sounds exciting, but will it really put us on the right track? It rests on certain important assumptions that are plausible but still largely unverified. Most crucially, is it true that changing minds is ultimately about changing connectomes? That's the obvious implication of theories that reduce perception, thought, and other mental phenomena to patterns of spiking generated by patterns of neural connections. Testing these theories would tell us whether connectionism really makes sense. It's a fact that the four R's of connectome change exist in the brain, but right now we can only speculate about how they're involved in learning. In the Darwinian view, synapses, branches, and neurons are created to endow the brain with new potential to learn. Some of this potential is actualized by Hebbian strengthening, which enables certain synapses, branches, and neurons to survive. The rest are eliminated to clear away unused potential. Without careful scrutiny of these theories, it's unlikely that we'll be able to harness the power of the four R's effectively.

To critically examine the ideas of connectionism, we must subject them to empirical investigation. Neuroscientists have danced around this challenge for over a century without having truly taken it on. The problem is that the doctrine's central quantity — the connectome — has been unobservable. It has been difficult or impossible to study the connections between neurons, because the methods of neuroanatomy have only been up to the coarser task of mapping the connections between brain regions.

We're getting there — but we have to speed up the process radically. It took over a dozen years to find the connectome of the worm

C. elegans, and finding connectomes in brains more like our own is of course much more difficult. In the next part of this book I'll explore the advanced technologies being invented for finding connectomes and consider how they'll be deployed in the new science of connectomics.

PART IV

CONNECTOMICS

EIGHT

Seeing Is Believing

SMELLING WHETS THE appetite, and listening saves relationships, but seeing is believing. More than any other sense, we trust our eyes to tell us what is real. Is this just a biological accident, the result of the particular way in which our sense organs and brains happened to evolve? If our dogs could share their thoughts by more than a bark or a wag of the tail, would they tell us that smelling is believing? As a bat dines on an insect, captured in the darkness of night by following the echoes of ultrasonic chirps, does it pause to think that hearing is believing?

Or perhaps our preference for vision is more fundamental than biology, based instead on the laws of physics. The straight lines of light rays, bent in an orderly fashion by a lens, preserve spatial relationships between the parts of an object. And images contain so much information that — until the development of computers — they could not easily be manipulated to create forgeries.

Whatever the reason, seeing has always been central to our beliefs. In the lives of many Christian saints, visions of God — apocalyptic or serene — often triggered the conversion of pagans into believers. Unlike religion, science is supposed to employ a method based on the formulation and empirical testing of hypotheses. But science, too, can be propelled by visual revelations, the sudden and simple sight of something amazing. Sometimes science is just seeing.

In this chapter I'll explore the instruments that neuroscientists have created to uncover a hidden reality. This might seem like a distraction from the real subject at hand — the brain — but I hope to convince you otherwise. Military historians dwell on the cunning gambits of daring generals, and the uneasy dance of soldiers and statesmen. Yet in the grand scheme of things, such tales may matter less than the backstory of technological innovation. Through the invention of the gun, the fighter plane, and the atomic bomb, weapon makers have repeatedly transformed the face of war more than any general ever did.

Historians of science likewise glorify great thinkers and their conceptual breakthroughs. Less heralded are the makers of scientific instruments, but their influence may be more profound. Many of the most important scientific discoveries followed directly on the heels of inventions. In the seventeenth century Galileo Galilei pioneered telescope design, increasing magnifying power from 3× to 30×. When he pointed his telescope at the planet Jupiter, he discovered moons orbiting around it, which overturned the conventional wisdom that all heavenly bodies circled the Earth.

In 1912 the physicist Lawrence Bragg showed how to use x-rays to determine the arrangement of atoms in a crystal, and three years later, at the tender age of twenty-five, he won the Nobel Prize for his work. Later on, x-ray crystallography enabled Rosalind Franklin, James Watson, and Francis Crick to discover the double-helix structure of DNA.

Have you heard the joke about two economists walking down the street? "Hey, there's a twenty-dollar bill lying on the sidewalk!" one of them says. "Don't be silly," says the other. "If there were, someone would have picked it up." The joke makes fun of the efficient market hypothesis (EMH), the controversial claim that there exists no fair and certain method of investment that can outperform the average return for a financial market. (Bear with me — you'll see the relevance soon.)

Of course, there are *uncertain* ways of beating the market. You can glance at a news story about a company, buy stock, and gloat when it goes up. But this is no more certain than a good night in Vegas. And there are unfair ways of beating the market. If you work for a pharmaceutical company, you might be the first to know that a drug is succeeding in clinical trials. But if you buy stock in your company based on such nonpublic information, you could be prosecuted for insider trading.

Neither of these methods fulfills the "fair" and "certain" criteria of the EMH, which makes the strong claim that no such method exists. Professional investors hate this claim, preferring to think they succeed by being smart. The EMH says that either they're lucky or they're unscrupulous.

The empirical evidence for and against the EMH is complex, but the theoretical justification is simple: If new information indicates that a stock will appreciate in value, then the first investors to know that information will bid the price up. And thus, says the EMH, there are no good investment opportunities available, just as there are never (well, *almost* never) twenty-dollar bills lying on the sidewalk.

What does this have to do with neuroscience? Here's another joke: "Hey, I just thought of a great experiment!" one scientist says. "Don't be silly," says the other. "If it were a great experiment, someone would already have done it." There's an element of truth to this exchange. The world of science is full of smart, hard-working people. Great experiments are like twenty-dollar bills on the sidewalk: With so many scientists on the prowl, there aren't many left. To formalize this claim, I'd like to propose the *efficient science hypothesis* (ESH): There exists no fair and certain method of doing science that can outperform the average.

How can a scientist make a truly great discovery? Alexander Fleming discovered and named penicillin after finding that one of his bacterial cultures had accidentally become contaminated by the fungus that produces the antibiotic. Breakthroughs like this are serendipitous.

If you want a more reliable method, it might be better to search for an "unfair" advantage. Technologies for observation and measurement might do the trick.

After hearing rumors of the invention of the telescope in Holland, Galileo quickly built one of his own. He experimented with different lenses, learning how to grind glass himself, and eventually managed to make the best telescopes in the world. These activities uniquely positioned him to make astronomical discoveries, because he could examine the heavens using a device others didn't have. If you're a scientist who purchases instruments, you could strive for better ones than your rivals by excelling at fundraising. But you'd gain a more decisive advantage by building an instrument that money can't buy.

Suppose you think of a great experiment. Has it already been done? Check the literature to find out. If no one has done it, you'd better think hard about why not. Maybe it's not such a great idea after all. But maybe it hasn't been done because the necessary technologies did not exist. If you happen to have access to the right machines, you might be able to do the experiment before anyone else.

My ESH explains why some scientists spend the bulk of their time developing new technologies rather than relying on those that they can purchase: They are trying to build their unfair advantage. In his 1620 treatise the *New Organon,* Francis Bacon wrote:

> It would be an unsound fancy and self-contradictory to expect that things which have never yet been done can be done except by means which have never yet been tried.

I would strengthen this dictum to:

> *Worthwhile things that have never yet been done can only be done by means that have never yet existed.*

It's at those moments when new means exist — when new technologies have been invented — that we see revolutions in science.

To find connectomes, we will have to create machines that produce clear images of neurons and synapses over a large field of view. This will be an important new chapter in the history of neuroscience, which is perhaps best seen not as a series of great ideas but as a series of great inventions, each of which surmounted a once insuperable barrier to observing the brain. It now seems trivial to say that the brain is made of neurons, but the path to this idea was tortuous. And for a simple reason—for a long time, it was impossible to see neurons.

Living sperm were first observed in 1677 by Antonie van Leeuwenhoek, a Dutch textile merchant turned scientist. Leeuwenhoek made the discovery with his homebrew microscope, but he didn't fully recognize its significance. He did not prove that the sperm, rather than the surrounding fluid in semen, were the agents of reproduction. And he had no inkling of the process of fertilization by which an egg and a sperm unite. But by paving the way to these discoveries by his successors, Leeuwenhoek's work was epoch-making.

Three years earlier, Leeuwenhoek had examined a droplet of lake water with his microscope. He saw tiny objects moving around and decided that they were alive. He called them "animalcules" and wrote a letter about them to the Royal Society of London. By now we are completely used to the idea of microscopic organisms and have difficulty imagining how much they must have stunned his contemporaries. At the time, though, Leeuwenhoek's claims seemed so fantastic that they provoked suspicions of fraud. To allay these fears, he sent the Royal Society testimonial letters from eight eyewitnesses, including three clergymen, a lawyer, and a physician. After several years his claims were finally vindicated, and the Society honored him with membership.

Leeuwenhoek is sometimes called the father of microbiology. This field turned out to have great practical significance in the nineteenth century, when scientists like Louis Pasteur and Robert Koch showed that microbial infection can cause disease. Microbiology was also critical in the development of the cell theory. This cornerstone of modern biology, formulated in the nineteenth century, holds that all organisms are composed of cells. Microscopic organisms are those that are composed of just a single cell.

Most of the members of the Royal Society were wealthy men with the time to devote themselves to intellectual pursuits. Leeuwenhoek was not born rich, but by age forty he had secured enough income to turn his attention to science. He did not study at a university, and did not know Latin or Greek. How did this self-educated man from humble origins achieve so much?

Leeuwenhoek did not invent the microscope; the credit goes to eyeglass craftsmen who worked at the end of the sixteenth century. Like today's microscopes, the first ones combined multiple lenses, but they could magnify only 20 to 50 times. Leeuwenhoek's microscopes delivered up to 10 times better magnification with a single, very powerful lens. We can't be certain how he learned to make such outstanding lenses, because he kept his methods secret. This was Leeuwenhoek's "unfair" advantage: He made better microscopes than the ones used by his rivals.

When Leeuwenhoek died, his methods were lost. Later, in the eighteenth century, technical improvements made the multilens ("compound") microscope more powerful than Leeuwenhoek's. Scientists were able to see the structures of plant and animal tissues more clearly, which resulted in the acceptance of the cell theory in the nineteenth century. Yet there was one place where the theory ran into trouble: the brain. Microscopists could see the cell bodies of neurons and the branches extending from them. But they lost track of the branches after a short distance. All they could see was a dense tangle, and no one knew what happened there.

The problem was resolved by a breakthrough in the second half of the nineteenth century. An Italian physician named Camillo Golgi invented a special method of staining brain tissue. Golgi's method stains only a few neurons; it leaves almost all of them unstained and hence invisible. Figure 26 may still look a little crowded, but we can make out the shapes of individual neurons. Golgi's scientific rival, the Spanish neuroanatomist Santiago Ramón y Cajal, was presumably viewing something like this in his microscope when he drew the image shown in Figure 1.

Golgi's new method was an extraordinary advance. To appreciate

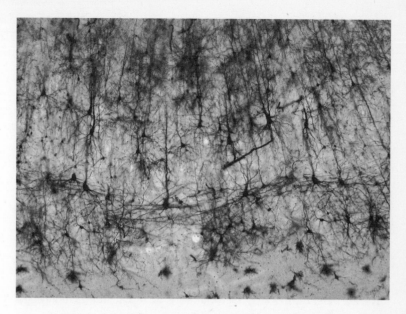

FIGURE 26. *Golgi staining of neurons in the cortex of a monkey*

why, let's imagine the branches of neurons as entangled strands of yellow spaghetti. (I introduced this metaphor earlier, but it's now even more apt, given Golgi's national origin.) Cooks with extremely bad eyesight see only a yellow mass on the plate, because the individual strands are too blurry to be distinguished. Now suppose that a single dark strand is mixed in with the others (see Figure 27, left). Even with blurry vision, it's possible to follow the path of the dark strand (right).

As an invention, a microscope might seem more glamorous than a stain. Its metal and glass parts are impressive, and can be designed using the laws of optics. A stain isn't much to look at; it might even smell bad. Stains are often discovered by chance rather than design. Actually, we still don't know why Golgi's stain marks only a small fraction of neurons. All we know is that it works. In any case, Golgi's stain and others have played an important role in the history of neuroscience. "The gain in the brain lies mainly in the stain," neuroanatomists like to say. Golgi's is simply the most famous.

Science can languish for a long time if the proper technology does

 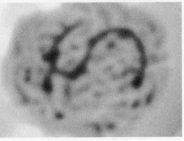

FIGURE 27. *Why Golgi staining works: photograph of pasta before* (left)
and after (right) *blurring*

not exist. Without the right kind of data, it can be impossible to make
progress, no matter how many smart people are working on the problem. The nineteenth-century struggle to see neurons lasted until the
invention of Golgi's staining method, which was soon used most av-
idly, and most illustriously, by Cajal. In 1906 Golgi and Cajal voyaged
to Stockholm to receive the Nobel Prize "in recognition of their work
on the structure of the nervous system." As is customary, both scientists gave special lectures describing their research. But rather than
celebrate their joint honor, the two men took the opportunity to attack
each other.

They had long been embroiled in a bitter dispute. Golgi's staining
method had finally revealed neurons to the world, but the limited resolution of the microscope still left ambiguities. When Cajal looked in
his microscope, he saw points at which two stained neurons contacted
each other but still remained separate. When Golgi looked in *his* microscope, he saw such points as locations where neurons had fused together into a continuous network, forming a kind of supercell.

By 1906 Cajal had convinced many of his contemporaries that a gap
existed, but it was still unclear how neurons could communicate with
each other if they were not physically continuous. Three decades later,
Otto Loewi and Sir Henry Dale were awarded the Nobel Prize "for
their discoveries relating to chemical transmission of nerve impulses."
They had found conclusive evidence that neurons can send messages
by secreting neurotransmitter molecules, and receive messages by

sensing them. The idea of a chemical synapse explained how two neurons could communicate across a narrow gap.

Still, no one had actually seen a synapse. In 1933 the German physicist Ernst Ruska built the first electron microscope, which used electrons rather than light to yield much sharper images. Ruska moved to Siemens and developed a commercial product. After World War II the electron microscope grew in popularity. Biologists learned how to cut their specimens into extremely thin slices and then image the slices. Finally, they had clear pictures.

The first images of synapses, obtained in the 1950s, showed that two neurons did not fuse at a synapse. There was definitely a boundary separating the two cells, and sometimes one could even see an extremely narrow gap between them. These are features that cannot be seen clearly with a light microscope, which is why Golgi and Cajal had been unable to resolve their debate.

With this new information, victory to Cajal. Or so it seemed. In the end, Golgi turned out to be right as well. In addition to chemical synapses, the brain also contains electrical synapses, as I mentioned earlier. At this type of synapse, special ion channels span the cleft between the two membranes and function as tunnels through which ions, electrically charged atoms, can travel from the inside of one neuron to the inside of another. An electrical synapse carries electrical signals directly between neurons without the need for an intermediate chemical signal, and effectively fuses two cells into one continuous supercell, as Golgi envisioned.

I've billed the electron microscope as the invention that enabled the imaging of synapses. But new stains were also crucial. With electron microscopy, it's sensible to use "dense" staining methods, which mark all neurons. The combination of electron microscope and dense stain revealed what neuroscientists had already imagined but had never seen clearly — the entangling of the branches of many neurons. While the Golgi stain showed the shape of a neuron, it gave the false impression that neurons are islands surrounded by large expanses of nothing. In fact, brain tissue is packed full of neurons and their branches, as you can see at the left in Figure 28. This image resembles what you'd see if

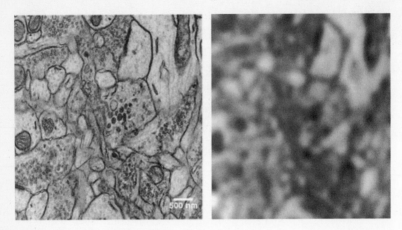

FIGURE 28. *Cross-sections of axons and dendrites imaged by an electron microscope, before* (left) *and after* (right) *blurring*

you cut through a tangled mass of spaghetti. The cut ends of the individual strands would have circular or elliptical cross-sections like the cross-sections of neural branches in the image.

The laws of physics limit the resolving power of a light microscope to the wavelength of light, which is a fraction of a micrometer. Details smaller than this are blurred, a barrier known as the diffraction limit. Shown at the right in Figure 28 is another version of the electron microscope image, this one artificially blurred to simulate how it would look in a light microscope. The cross-sections of the thinnest branches of neurons are no longer clearly visible. That's why sparse methods of staining like Golgi's, which mark only a few neurons, were necessary when using the light microscope. The electron microscope's much higher resolving power makes it possible to see all neurons at the same time with dense staining.

An electron microscopic image shows only two-dimensional cross-sections of neurons, however. To see neurons in their full glory, we need three-dimensional images. These can be obtained by slicing up brain tissue using a high-tech version of the machine in your local delicatessen, and then imaging every slice. Cutting might sound trivial, but the slices must be tens of thousands of times thinner than your typical prosciutto. For this, we need a most unusual knife.

I have always had a fetish for knives. When I was a Cub Scout I got my first pocketknife, with two cheap blades that quickly tarnished. An older boy showed me his bright red Swiss Army knife bristling with shiny tools as well as blades, and I was overcome with envy. Today I prefer German chef's knives made of carbon stainless steel. (I am not enough of a fanatic to prefer the sharper knives that rust.) I love the whirring sound of knife edge on sharpening steel, and the satisfying feeling of gliding through the flesh of a tomato.

Diamonds, on the other hand, I never understood. Yes, they sparkle, but so does a piece of cubic zirconium or even cut glass. How much more lovely is the pale blue of aquamarine or the blood red of ruby! Surely those beautiful colors are more passionate than the vacuous transparency of diamond.

But then I met the diamond knife.

To understand how special this tool is, let's start with a riddle: What is the difference between a knife and a saw? You might reply that a saw is jagged while a knife edge is smooth, or that a knife is tapered to a sharp edge while the edge of a saw is blunt. But these distinctions melt away under a microscope. However smooth it may appear to the naked eye, the edge of any metal knife looks blunt and jagged when magnified. Even the finely honed blade of a sushi chef looks as crude as a bludgeon.

But there is one knife whose rarefied perfection holds up to close inspection. The edge of a well-honed diamond knife looks perfectly sharp and smooth, even under an electron microscope. It is just 2 nanometers wide, or about 12 carbon atoms. Small nicks on the atomic scale may be visible, but these are rare on a high-quality blade. Its superiority to a metal knife is obvious in the electron microscope images shown in Figure 29.

The diamond knife is the most advanced of the many types of blades used during the centuries-long history of microscopy. The cel-

 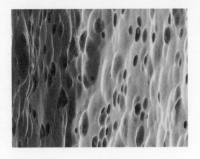

FIGURE 29. *Knives: diamond* (left) *versus metal* (right)

lular structures of plant and animal tissues are best viewed in specimens prepared by cutting slices, which, for light microscopy, should be as thin as a human hair. At first, specimens were prepared manually using razor blades. In the nineteenth century, inventors developed machines called microtomes. The piece of tissue advanced toward the knife (or vice versa) in small steps, yielding uniformly thin slices.

A microtome can cut as thin as a few micrometers. This was more than sufficient for light microscopy, but the invention of the electron microscope made it necessary to cut even thinner. Keith Porter and Joseph Blum constructed the first ultramicrotomes in 1953. These machines cut slices that were an astounding 50 nanometers in thickness, more than a thousand times thinner than a human hair. Ultramicrotomes were first fitted with glass knives, but diamond knives proved better. Their perfect sharpness yields clean cuts, and they are durable enough to cut many slices before dulling. As you might imagine, brain tissue has to be prepared very carefully before it's cut in an ultramicrotome. Because of its soft, tofu-like consistency, the tissue would fall apart if it were sliced fresh, so it's embedded in an epoxy resin that hardens into a plastic block.

Ultramicrotomes were first used to obtain single two-dimensional images, like those shown in the figures in this chapter. In the 1960s researchers took the obvious next step of imaging a long series of many slices. This method, known as *serial electron microscopy,* produced a three-dimensional image by stacking up two-dimensional images of many slices. In principle, it's possible to image the entirety of every neuron and synapse in a piece of brain tissue, and perhaps even an entire brain. That's just what we need for finding connectomes. In practice, however, the method is laborious. Because the slices are so fragile, it's not easy to pick them up and place them inside the electron microscope. Every now and then a slice is damaged or lost. There is ample opportunity for error, as even a small piece of brain produces a huge number of thin slices.

For several decades we had no solution to this problem. And then a German physicist had a simple but brilliant idea.

Heidelberg, a lovely German city about an hour's drive from Frankfurt, seems an unlikely incubator for futuristic technologies. A half-ruined castle draws tourists in droves. The old part of town is paved with cobblestones and peppered with bars and restaurants serving raucous students from Ruprechts Karl University. If you're feeling the need to think profound thoughts, head to the Philosopher's Walk, a mountain trail with a splendid view of the Neckar River, where you can channel the spirits of Heidelberg intellectuals like the philosophers Hegel and Hannah Arendt.

Near one of the Neckar's bridges sits a brick building, the Max Planck Institute for Medical Research, at 29 Jahnstrasse. The building appears modest, but it has housed five Nobel laureates in its history. It is one of eighty elite institutes operated by the Max Planck Society, the crown jewel of German science. Each institute is run by several directors, each with a large budget, a small army of research assistants, and a skilled technical staff. The decisions of the Max Planck Society are made by vote of its members, the several hundred directors of the institutes. It is a very exclusive club.

One of the previous directors at 29 Jahnstrasse was Bert Sakmann, who shared the Nobel Prize for his invention of patch clamp recording, now one of the standard tools of the neurophysiologist. He recruited the physicist Winfried Denk to become a new director at the institute.

Denk is a large man, with the commanding physical presence of a German feudal lord. (Maybe that's not surprising; Max Planck directors are about as close as you can get to feudal lords in the modern world.) Denk also impresses with his wit. A science lab is no magnet for comic geniuses, but there are some exceptions. I'll never forget the seminar of a brilliant applied mathematician filled with hilarious riffs on sex, drugs, and rock and roll, which made me laugh so hard that my stomach hurt and tears rolled down my face, obscuring my view of the equations. Denk's one-liners reveal his agile mind, although to get the full effect you must be a night owl, as he prefers the "Dracula schedule" of rising late and working until close to dawn. The experience is well worth it, as the quips and aphorisms flow most freely after midnight.

In the basement of 29 Jahnstrasse, three electron microscopes are protected from temperature fluctuations by special enclosures. Pumps evacuate their metal chambers so that electrons may fly freely inside without colliding with air molecules. The microscopes are a bit finicky — at any given time, one might be in need of repair. But the others image brain tissue continuously for weeks or months.

When Denk first arrived in Heidelberg, he was already world famous as one of the inventors of the two-photon microscope. (I mentioned earlier that this tool was used to see creation and elimination of synapses in living brains of animals.) After shaking up light microscopy, he decided to automate serial electron microscopy. His idea was simple: repeatedly image the face of the specimen exposed by cutting, rather than image the slices.

In 2004 Denk unveiled his invention, an automated system consisting of an ultramicrotome mounted inside the vacuum chamber of an electron microscope. He called his method SBFSEM, short for "serial block face scanning electron microscopy." By bouncing electrons off a block of brain tissue, it's possible to acquire a 2D image of the block face. Then the blade of the ultramicrotome scrapes a thin slice off the block, exposing a new face, which is imaged again. This process is repeated to acquire a stack of 2D images similar to those produced by conventional serial electron microscopy.

Why is it better to image the block face rather than the slices? Because the block is sturdy while the slices are fragile. Even if slices are not lost through mishandling, each ends up slightly deformed in a different way. Stacking up the images of the slices produces a corrupted 3D image. In contrast, images of the block face contain little or no distortion because the sturdy block deforms so little.

Imaging the block face allowed the ultramicrotome to be placed inside the electron microscope, creating an automated system that integrated both cutting and imaging. This improved reliability by eliminating the error-prone manual transfer of slices from ultramicrotome to microscope. The slices were as thin as 25 nanometers, half the 50 nanometers feasible with manual cutting and collection.

Like mountain climbers, scientists strive to be first. Glory goes to discoverers, not followers. But science also resembles investing — you can be too early rather than too late. In his 2004 paper Denk acknowledged an earlier inventor named Stephen Leighton, who had developed a similar idea in 1981. Leighton's invention was too early to be practical, as it would have produced more data than anyone could handle at the time. By the time Denk came up with the idea independently, computers had advanced enough to store the large amounts of data.

How do you know when an idea's time has really come? As with investments, it often becomes obvious only in retrospect, when less upside remains. Simultaneous invention by two people is one sign, but even more telling is invention of two different solutions to the same problem. As it happens, Denk's work was paralleled by another effort to automate the process of seeing smaller.

No ivy grows on the Northwest Building at Harvard University. The smooth glass exterior exudes no hint of history, which is appropriate for a building that houses the cutting edge of scientific research at Harvard. Enter the expansive lobby and wander down to a basement room. Before your eyes is a bewildering machine — a complex, Rube Goldberg contraption (see Figure 30). It's not clear what to look at, un-

FIGURE 30. *The Harvard ultramicrotome*

til the slow motions of a tiny plastic block catch your eye. Its transparency emits a hint of orange and envelops a black fleck, a stained piece of mouse brain.

Some other parts rotate lazily. Plastic tape is rolling off one reel and collecting on another, in the style of a 1970s reel-to-reel tape recorder. You spy yet another reel lying on the table beside the machine. Unrolling some tape, you hold it up to the light to see brain slices spaced at regular intervals along its length. Finally you realize that the function of the machine is to transform a piece of brain into something like a film strip, by cutting and collecting slice after slice on the tape.

Cutting slices is challenging enough; the problem of collecting them is even worse. As any amateur chef knows, thin slices often stick to the knife instead of falling onto the cutting board in an orderly way. The conventional ultramicrotome solves this problem using a water trough. The knife is mounted on one of the edges of the trough, so that the cut slices spread out nicely onto the surface of the water inside. The operator then plucks the slices out of the water one by one and takes them to the electron microscope for imaging. Mishandling during this process can result in annoying folds or in loss of entire slices.

The ultramicrotome at Harvard, like the conventional ones, uses a water trough to pull the ribbon of brain slices off the knife. The new element in the Harvard apparatus is a plastic tape, which ascends from the water's surface like a conveyor belt. (See the bottom of the image shown in Figure 31 for the plastic tape. You may be able to make out two slices of mouse brain touching end to end in the vertical ribbon centered above the tape.) Each slice sticks to the moving tape and is carried up out of the water into air, where it dries rapidly. The end result is a set of fragile slices stuck onto a much thicker and stronger tape that's collected on a reel. The important feature is that there's no possibility of human error, since the operator never needs to handle a slice manually. And the plastic tape is robust — virtually indestructible.

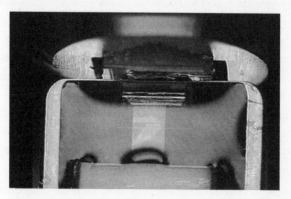

FIGURE 31. *Freshly cut brain slices being collected by a plastic tape rising out of the water*

The first prototype of ATUM, the automated tape-collecting ultramicrotome, was built in more modest surroundings — a garage thousands of miles away in the city of Alhambra, near Los Angeles. Its inventor, Ken Hayworth, is tall, thin, and bespectacled, with a determined walk and an intense way of talking. As an engineer at NASA's Jet Propulsion Laboratory, Hayworth built inertial guidance systems for spacecraft. Then he switched careers, enrolling in a neuroscience Ph.D. program at the University of Southern California. Hayworth has a lot of energy, which may explain why he used his spare time to build a new machine for slicing brains in his garage.

The prototype sliced at 10 microns, too thick for electron microscopy, but it demonstrated the basic idea. One day Hayworth received a phone call from out of the blue. It was Jeff Lichtman, the Harvard expert on synapse elimination, calling to suggest a collaboration. Hayworth set up shop at Harvard, and built another ATUM that was capable of cutting at 50 nanometers, the thickness achieved by a conventional ultramicrotome. Lichtman egged him on, and the machine eventually achieved 30 nanometers. To image the slices, Hayworth teamed up with Narayanan "Bobby" Kasthuri. The two made an entertainingly odd couple. Other lab members joked that Kasthuri *seemed* crazy, with his wild hair and even wilder stories, but that Hayworth was actually the crazy one. (More on this inside joke later.) They and another researcher, Richard Schalek, employed a scanning electron microscope for imaging, the same instrument modified by Denk.

Denk's invention eliminates the need to collect sections; Hayworth's makes collecting them reliable. Other inventors are working on their own schemes to improve cutting and imaging. For example, Graham Knott has shown how to use a beam of ions to vaporize the top few nanometers off the top of the block. This technique is similar to Denk's but eliminates the need for a diamond knife. Such inventions are just the beginning of what I anticipate will be a golden age in serial electron microscopy.

Along with that golden age comes a new challenge for neuroscience, the era of too much information. Just one cubic millimeter of brain tissue could yield a petabyte of image data. This is equivalent to a digital photo album containing one billion images. An entire mouse brain is a thousand times larger, a human brain a thousand times larger still. So the improvements in cutting, collecting, and imaging are not by themselves enough to find connectomes. Imaging every neuron and synapse will produce torrents of information, overwhelming the ability of any human to comprehend. To find connectomes, we need not only machines for *making* images, but also machines for *seeing* them.

NINE

Following the Trail

T HE ANCIENT GREEKS told the story of King Minos, who kept a beautiful white bull for himself instead of offering it as a sacrifice. The gods, angry at his greed, punished Minos by driving his wife mad with lust for the bull. She gave birth to the Minotaur, a monster with two legs and two horns. Minos imprisoned her deadly offspring in the Labyrinth, a mazelike structure ingeniously constructed by the great engineer Daedalus. Eventually the hero Theseus came from Athens and killed the Minotaur. To find his way back out of the Labyrinth, he followed a thread supplied by his lover Ariadne, the daughter of Minos.

Connectomics reminds me of this myth. Like the Labyrinth, the brain must deal with the consequences of destructive emotions such as greed and lust, while also inspiring acts of ingenuity and love. Try to imagine yourself traveling through the axons and dendrites of the brain, like Theseus navigating the twisting passages of the Labyrinth. Perhaps you are a protein molecule sitting on a molecular motor car running on a molecular track. You are being transported on the long journey from your birthplace, the cell body, to your destination, the outer reaches of the axon. You patiently sit and watch as the walls of the axon go by.

If this journey sounds intriguing, let me invite you to embark on a virtual version. You will travel through images of the brain, rather than the brain itself. You'll trace the path of an axon or dendrite through a stack of images collected by the machines described in Chapter 8. It's a task essential for finding connectomes. In order to map the brain's connections, you have to see which neurons are connected by synapses, and you can't do it without knowing where the "wires" go.

To find an entire connectome, though, you'd have to explore every passage in the brain's labyrinth. To map just one cubic millimeter, you'd have to travel through miles of neurites and wade through a petabyte of images. Such laborious and careful analysis would be essential; a mere glance at the images would tell you nothing. This style of science seems far removed from Galileo's sighting of the moons of Jupiter or Leeuwenhoek's glimpse of sperm.

Today, our notion of "science as seeing" is being stretched to the limit by current technologies. No single person can possibly comprehend all the images now being collected by automated instruments. But if technology created the problem, maybe it can also solve it. Perhaps computers could trace the paths of all those axons and dendrites through the images. If our machines did most of the work for us, we'd be able to see connectomes.

The problem of dealing with huge quantities of data is not unique to connectomics. The world's largest scientific project is the Large Hadron Collider (LHC), a circular tube constructed one hundred meters underground, inside a twenty-seven-kilometer-long tunnel between Lake Geneva and the Jura Mountains. The LHC accelerates protons to great speeds and smashes them together to probe the forces between elementary particles. At one location on its circumference sits a gigantic apparatus called the Compact Muon Solenoid. It's designed to detect one billion collisions per second, of which one hundred are selected by computers that automatically sift through the data. Only these interesting events are recorded, but the data still flows at a torrential rate, as each event yields over one megabyte. The data is shipped to a network of supercomputers around the world for analysis.

To find entire connectomes of mammalian brains, we will need microscopes that produce images at data rates greater than those of the LHC. Can we analyze the data quickly enough to keep up? The scientists who compiled the *C. elegans* connectome encountered a similar challenge. To their surprise, it took more effort to analyze the images than to collect them.

In the mid-1960s, the South African biologist Sydney Brenner saw the possibility of using serial electron microscopy to map all the connections in a small nervous system. The term *connectome* had not been invented yet, and Brenner called the task "reconstruction of a nervous system." Brenner was working at the MRC Laboratory for Molecular Biology in Cambridge, England. At that time, he and others at the lab were establishing *C. elegans* as a standard animal for research on genetics. It later became the first animal to have its genome sequenced, and thousands of biologists study *C. elegans* today.

Brenner thought that *C. elegans* might also help us understand the biological basis of behavior. It did the standard things like feeding, mating, and laying eggs. It also gave canned responses to certain stimuli. For example, if you touched its head, it would recoil and swim away. Now suppose you found a worm that was incapable of one of these standard behaviors. If its offspring inherited the same problem, you could assume that the cause was a genetic defect, and try to pinpoint it. That kind of research would elucidate the relationship between genes and behaviors, which would already be valuable. But one could raise the stakes even further by examining the nervous systems of such mutant worms. Perhaps one would be able to identify particular neurons or pathways disrupted by the faulty gene. The prospect of studying the worm at all these levels — genes, neurons, and behavior — sounded truly exciting. But the whole plan hinged on something that Brenner did not have: a map of the normal worm's nervous system. Without that, it would be difficult to discern what was different about the nervous systems of mutants.

Brenner was aware of the early twentieth-century attempt of Richard Goldschmidt, a German-American biologist, to map the nervous system of another species of worm, *Ascaris lumbricoides.* Goldschmidt's light microscope did not have enough resolution to show the branches of neurons clearly, or reveal synapses. Brenner decided to try something similar with *C. elegans,* but using the superior technology of the electron microscope and the ultramicrotome.

C. elegans is just one millimeter long, much smaller than *Ascaris,* which can grow up to a foot in the intestines of its human hosts. Converting the entire *C. elegans* worm, like a tiny sausage, into slices thin enough for electron microscopy could be accomplished with a mere several thousand cuts. Nichol Thomson, a member of Brenner's team, found it impossible to slice up an entire worm without error, owing to the technical difficulties of the not-yet-automated slicing process, but he could manage a large fraction of a worm. Brenner decided to combine images from segments of several worms. It was a reasonable strategy because the worm's nervous system is so standardized.

Thomson sliced up worms until he had covered every region of the worm's body at least once. The slices were placed one by one in an electron microscope and imaged (see Figure 32). This laborious process eventually yielded a stack of images representing the entire nervous system of *C. elegans.* All of the worm's synapses were there.

You might think Brenner and his team were done at that point. Isn't a connectome just the entirety of all synapses? In fact, they had only just begun. Although the synapses were all visible, their organization was still hidden. In effect, the researchers had collected a jumbled-up bag of synapses. To find the connectome, they needed to sort out which synapses belonged to which neurons. They couldn't tell from a single image, which showed only two-dimensional cross-sections of neurons. But if they could follow the successive cross-sections of a single neuron through a sequence of images, they could determine which synapses belonged to it. And if this could be done for all the neu-

FIGURE 32. *A slice of* C. elegans

rons, then the connectome would be found. In other words, Brenner's team would know which neurons were connected to which other neurons.

Again, think of a worm as a tiny sausage. But imagine this time that the sausage is stuffed with spaghetti. These spaghetti strands are its neurons, and our task is to trace the path of each one. Since we don't have x-ray vision, we ask the butcher to cut the sausage into many thin slices. Then we lay all the slices flat and trace each strand by matching its cut pieces from slice to slice.

To have any hope of tracing without errors, the slices must be extremely thin, less than the diameter of a spaghetti strand. Similarly, the slices of *C. elegans* had to be thinner than the branches of neurons, which can be less than 100 nanometers in diameter. Nichol Thomson cut slices about 50 nanometers thick — just thin enough to allow most branches of neurons to be traced reliably.

John White, who was trained as an electrical engineer, attempted to computerize the analysis of the images, but the technology was too primitive. White and a technician named Eileen Southgate had to resort to manual analysis. Cross-sections of the same neuron were marked with the same number or letter, as shown in the two images in Figure 33. To trace a single neuron in its entirety, the researchers repeatedly wrote the same symbol on the appropriate cross-section in successive images, like Theseus unrolling Ariadne's thread in the Labyrinth. Once the paths of neurons were traced, they went back to each synapse and noted the letters or numbers of the neurons involved in it. And in this way the *C. elegans* connectome slowly emerged.

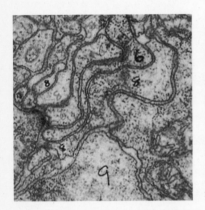 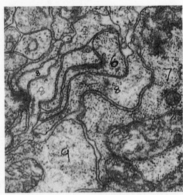

FIGURE 33. *Tracing the branches of neurons by matching their cross-sections in successive slices*

In 1986 Brenner's team published the connectome as an entire issue of the *Philosophical Transactions of the Royal Society of London*, a journal of the same society that had welcomed Leeuwenhoek as a member centuries before. The paper was titled "The Structure of the Nervous System of the Nematode *Caenorhabditis elegans*," but its running head was the pithier "The Mind of a Worm." The body of the text is a 62-page appetizer. The main course is 277 pages of appendices, which describe the 302 neurons of the worm along with their synaptic connections.

As Brenner had hoped, the *C. elegans* connectome turned out to be useful for understanding the neural basis of the worm's behaviors. For example, it helped identify the neural pathways important for behaviors like swimming away from a touch to the head. But only a small fraction of Brenner's original ambitions were realized. It wasn't for lack of images; Nichol Thomson had gathered plenty of them, from many worms. He had actually imaged worms with many types of genetic defects, but it was too laborious to analyze the images to detect the hypothesized abnormalities in their connectomes. Brenner had started out wanting to investigate the hypothesis that the "minds" of worms differ because their connectomes differ, but he had been unable to do so because his team had found only a single connectome, that of a normal worm.

Finding even one connectome was by itself a monumental feat. Analyzing the images consumed over a dozen years of effort in the 1970s and 1980s — much more labor than was required to cut and image the slices. David Hall, another *C. elegans* pioneer, has made these images available online in a fascinating repository of information about the worm. (The vast majority of them remain unanalyzed today.) The toil of Brenner's team served as a cautionary note, effectively warning other scientists, "Don't try this at home."

The situation began to improve in the 1990s, when computers became cheaper and more powerful. John Fiala and Kristen Harris created a software program that facilitated the manual reconstruction of the shapes of neurons. The computer displayed images on a screen and allowed a human operator to draw lines on top of them using a mouse. This basic functionality, familiar to anyone who has used computers to create drawings, was then extended to allow a person to trace a neuron through a stack of images, drawing a boundary around each cross-section. As the operator worked, each image in the stack would become covered with many boundary drawings. The computer kept track of all the cross-section boundaries that belonged to each neuron, and displayed the results of the operator's labors by coloring within the lines. Each neuron was filled with a different color, so that the stack of images resembled a three-dimensional coloring book. The computer could also render parts of neurites in three dimensions, as in the image shown in Figure 34.

FIGURE 34. *Three-dimensional rendering of neurite fragments reconstructed by hand*

With this process, scientists could do their work much more efficiently than Brenner's team had in the *C. elegans* project. Images were
now stored neatly on the computer, so researchers no longer had to
deal with thousands of photographic plates. And using a mouse was
less cumbersome than manual marking with felt-tip pens. Nevertheless, analyzing the images still required human intelligence and was
still extremely time-consuming. Using their software to reconstruct
tiny pieces of the hippocampus and the neocortex, Kristen Harris and
her colleagues discovered many interesting facts about axons and dendrites. The pieces were so small, however, that they contained only
minuscule fragments of neurons. There was no way to use them to
find connectomes.

Based on the experience of these researchers, we can extrapolate that manual reconstruction of just one cubic millimeter of cortex could take a million person-years, much longer than it would take
to collect the electron microscopic images. Because of these daunting
numbers, it's clear that the future of connectomics hinges on automat-
ing image analysis.

Ideally we'd have a computer, rather than a person, draw the boundaries of each neuron. Surprisingly, though, today's computers are not
very good at detecting boundaries, even some that look completely

obvious to us. In fact, computers are not so good at any visual task. Robots in science fiction movies routinely look around and recognize the objects in a scene, but researchers in artificial intelligence (AI) are still struggling to give computers even rudimentary visual powers.

In the 1960s researchers hooked up cameras to computers and attempted to build the first artificial vision systems. They tried to program a computer to turn an image into a line drawing, something any cartoonist could do. They figured it would be easy to recognize the objects in the drawing based on the shape of their boundaries. It was then that they realized how bad computers are at seeing edges. Even if the images were restricted only to stacks of children's blocks, it was challenging for the computers to detect the boundaries of the blocks.

Why is this task so difficult for computers? Some subtleties of boundary detection are revealed by a well-known illusion called the Kanizsa triangle (Figure 35). Most people see a white triangle superimposed on a black-outlined triangle and three black circles. But it's arguable that the white triangle is illusory. If you look at one of its corners while blocking the rest of the image with your hand, you'll see a partially eaten pie (or a Pac-Man, if you remember that video game from the 1980s) rather than a black circle. If you look at one of the V's while blocking the rest of the image with both hands, you won't see any boundary where you used to see a side of the white triangle. That's because most of the length of each side is the same color as the back-

FIGURE 35. *The "illusory contours" of the Kanizsa triangle*

ground, with no jump in brightness. Your mind fills in the missing parts of the sides—and perceives the superimposed triangle—only when provided with the context of the other shapes.

This illusion might seem too artificial to be important for normal vision. But even in images of real objects, context turns out to be essential for the accurate perception of boundaries. The first panel of Figure 36, a zoomed-in view of part of an electron microscope image of neurons, shows little evidence of a boundary. As subsequent panels reveal more of the surrounding pixels, a boundary at the center becomes evident. Detecting the boundary leads to the correct interpretation of the image (next-to-last panel); missing the boundary would lead to an erroneous merger of two neurites (last panel). This kind of mistake, called a merge error, is like a child's use of the same crayon to color two adjacent regions in a coloring book. A split error (not shown) is like the use of two different crayons to color a single region.

Granted, this sort of ambiguity is relatively rare. The one shown in the figure presumably arose because the stain failed to penetrate one location in the tissue. In most of the rest of the image, however, it would be obvious whether or not there is a boundary even in a zoomed-in view. Computers are able to detect boundaries accurately at these easy locations but still stumble at a few difficult ones, because they are less adept than humans at using contextual information.

Boundary detection is not the only visual task that computers need to perform better if we want to find connectomes. Another task involves recognition. Many digital cameras are now smart enough to locate and focus on the faces in a scene. But sometimes they erroneously

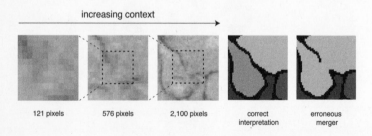

FIGURE 36. *The importance of context for boundary detection*

focus on some object in the background, showing that they still don't recognize faces as well as people do. In connectomics, we'd like computers to perform a similar task, and to do it flawlessly: look through a set of images and find all the synapses.

Why have we failed (so far) to create computers that see as well as humans? In my view, it is because we see so well. The early AI researchers focused on duplicating capabilities that demand great effort from humans, such as playing chess or proving mathematical theorems. Surprisingly, these capabilities ended up being not so difficult for computers — in 1997 IBM's Deep Blue supercomputer defeated the world chess champion Garry Kasparov. Compared with chess, vision seems childishly simple: We open our eyes and instantly see the world around us. Perhaps because of this effortlessness, early AI researchers didn't anticipate that vision would be so difficult for machines.

Sometimes the people who are the best at doing something are the worst teachers. They themselves can do the task unconsciously, without thinking, and if they're asked to explain what they do, they have no idea. We are all virtuosos at vision. We've always been able to do it, and we can't understand an entity that can't. For these reasons we're lousy at teaching vision. Luckily we never have to, except when our students are computers.

In recent years some researchers have given up on instructing computers to see. Why not let a computer teach itself? Collect a huge number of examples of a visual task performed by humans, and then program a computer to imitate these examples. If the computer succeeds, it will have "learned" the task without any explicit instruction. This method is known as machine learning, and it's an important subfield of computer science. It has yielded the digital cameras that focus on faces, as well as many other successes in AI.

Several laboratories around the world, including my own, are using machine learning to train computers to see neurons. We start by making use of the kind of software that John Fiala and Kristen Harris developed. People manually reconstruct the shapes of neurons, which serve as examples for the computer to emulate. Viren Jain and Srini Turaga, my doctoral students at the time we began the work, have devised methods for numerically "grading" a computer's performance by measuring its disagreement with humans. The computer learns to see the shapes of neurons by optimizing its "grade" on the examples. Once trained in this way, the computer is given the task of analyzing images that humans have not manually reconstructed. Figure 37 shows a computer reconstruction of retinal neurons. This approach, though still in its beginning stages, has already attained unprecedented accuracy.

FIGURE 37. *Neurons of the retina reconstructed automatically by computer*

Even with these improvements, the computer still makes errors. I'm confident that the application of machine learning will continue to reduce the error rate. But as the field of connectomics develops, computers will be called upon to analyze larger and larger images, and the absolute number of errors will remain large, even if the error *rate* is decreasing. In the foreseeable future, image analysis will never be 100 percent automatic — we will always need some element of human intelligence — but the process will speed up considerably.

• • •

It was the legendary inventor Doug Engelbart who first developed the idea of interacting with computers through a mouse. The full implications were not realized until the 1980s, when the personal-computer revolution swept the world. But Engelbart invented the mouse back in 1963, while directing a research team at the Stanford Research Institute, a California think tank. That same year, Marvin Minsky cofounded the Artificial Intelligence Laboratory (AI Lab) on the other side of the country, at the Massachusetts Institute of Technology. His researchers were among the first to confront the problem of making computers see.

Old-time computer hackers like to tell the story, perhaps apocryphal, of a meeting between these two great minds. Minsky proudly proclaimed, "We're going to make machines intelligent! We're going to make them walk and talk! We're going to make them conscious!" Engelbart shot back, "You're going to do that for computers? Well, what are you going to do for people?"

Engelbart laid out his ideas in a manifesto called "Augmenting Human Intellect," which defined a field he called Intelligence Amplification, or IA. Its goal was subtly different from that of AI. Minsky aimed to make machines smarter; Engelbart wanted machines that made *people* smarter.

My laboratory's research on machine learning belonged to the domain of AI, while the software program of Fiala and Harris was a direct descendant of Engelbart's ideas. It was not AI, as it was not smart enough to see boundaries by itself. Instead it amplified human intelligence, helping humans analyze electron microscopic images more efficiently. The field of IA is becoming increasingly important for science, now that it's possible to "crowdsource" tasks to large numbers of people over the Internet. The Galaxy Zoo project, for instance, invites members of the public to help astronomers classify galaxies by their appearance in telescope images.

But AI and IA are not actually in competition, because the best approach is to combine them, and that is what my laboratory is currently doing. AI should be part of any IA system. AI should take care of the easy decisions and leave the difficult ones for humans. The best way of making humans effective is to minimize the time they waste on trivial tasks. And an IA system is the perfect platform for collecting examples that can be used to improve AI by machine learning. The marriage of IA and AI leads to a system that gets smarter over time, amplifying human intelligence by a greater and greater factor.

People are sometimes frightened by the prospect of AI, having seen too many science fiction movies in which machines have rendered humans obsolete. And researchers can be distracted by the promise of AI, struggling in vain to fully automate tasks that would be more efficiently accomplished by the cooperation of computers and humans. That's why we should never forget that the ultimate goal is IA, not AI. Engelbart's message still comes through loud and clear for the computational challenges of connectomics.

These advances in image analysis are exciting and encouraging, but how quickly can we expect connectomics to progress in the future? We have all experienced incredible technological progress in our own lifetimes, especially in the area of computers. The heart of a desktop computer is a silicon chip called a microprocessor. The first microprocessors, released in 1971, contained just a few thousand transistors. Since then, semiconductor companies have been locked in a race to pack more and more transistors onto a chip. The pace of progress has been breathtaking. The cost per transistor has halved every two years, or — another way of looking at it — the number of transistors in a microprocessor of fixed cost has doubled every two years.

Sustained, regular doubling is an example of a type of growth called "exponential," after the mathematical function that behaves in this way. Exponential growth in the complexity of computer chips is known as Moore's Law, because it was foreseen by Gordon Moore in a 1965 article in *Electronics* magazine. This was three years before he helped found Intel, now the world's largest manufacturer of microprocessors.

The exponential rate of progress makes the computer business unlike almost any other. Many years after his prediction was confirmed, Moore quipped, "If the automobile industry advanced as rapidly as the semiconductor industry, a Rolls Royce would get half a million miles per gallon, and it would be cheaper to throw it away than to park it."

We are persuaded to throw our computers away every few years and buy new ones. This is usually not because the old computers are broken, but because they have been rendered obsolete.

Interestingly, genomics has also progressed at an exponential rate, more like semiconductors than automobiles. In fact, genomics has bounded ahead even faster than computers. The cost per letter of the DNA sequence has halved even faster than the cost per transistor.

Will connectomics be like genomics, with exponential progress? In the long run, it's arguable that computational power will be the primary constraint on finding connectomes. After all, the image analysis took up far more time than the image acquisition in the *C. elegans* project. In other words, connectomics will ride on the back of the computer industry. If Moore's Law continues to hold, then connectomics will experience exponential growth — but no one knows for sure whether that will happen. On the one hand, growth in the number of transistors in a single microprocessor has started to falter, a sign that Moore's Law could break down soon. On the other, growth might be maintained — or even accelerated — by the introduction of new computing architectures or nanoelectronics.

If connectomics experiences sustained exponential progress, then finding entire human connectomes will become easy well before the end of the twenty-first century. At present, my colleagues and I are preoccupied with surmounting the technical barriers to seeing connectomes. But what happens when we succeed? What will we do with them? In the next few chapters I'll explore some of the exciting possibilities, which include creating better maps of the brain, uncovering the secrets of memory, zeroing in on the fundamental causes of brain disorders, and even using connectomes to find new ways of treating them.

TEN

Carving

ONE DAY WHEN I was a boy, my father brought home a globe. I ran my fingers over the raised relief and felt the bumpiness of the Himalayas. I clicked the rocker switch on the cord and lay in bed gazing at the globe's luminous roundness in my darkened room. Later on, I was fascinated by a large folio book, my father's atlas of the world. I used to smell the leathery cover and leaf through the pages looking at the exotic names of faraway countries and oceans. My schoolteachers taught me and my classmates about the Mercator projection, and we giggled at the grotesque enlargement of Greenland with the same perverse enjoyment that came from a funhouse mirror or a newspaper cartoon on a piece of Silly Putty.

Today maps are practical items for me, not magical objects. As my childhood memories grow faint, I wonder whether my fascination with maps helped me cope with a fear of the world's vastness. Back then, I never ventured beyond the streets of my neighborhood without my parents; the city beyond seemed frightening. Placing the entire world on a sphere or enclosing it within the pages of a book made it seem finite and harmless.

In ancient times, fear of the world's vastness was not limited to children. When medieval cartographers drew maps, they did not leave unknown areas blank; they filled them with sea serpents, other imaginary monsters, and the words "Here be dragons." As the centuries passed, explorers crossed every ocean and climbed every mountain, gradually filling in the blank areas of the map with real lands. Today we marvel at photos of the Earth's beauty taken from outer space, and our communications networks have created a global village. The world has grown small.

Unlike the world, the brain seemed compact at first, fitting nicely inside the confines of a skull. But the more we know about the brain with its billions of neurons, the more it seems intimidatingly vast. The first neuroscientists carved the brain into regions and gave each one a name or number, as Brodmann did with his map of the cortex. Finding this approach too coarse, Cajal pioneered another one, coping with the immensity of the brain forest by classifying its trees like a botanist. Cajal was a "neuron collector."

We learned earlier why it's important to carve the brain into regions. Neurologists interpret the symptoms of brain damage using Brodmann's map. Every cortical area is associated with a specific mental ability, such as understanding or speaking words, and damage to that area impairs that particular ability. But why is it important to carve the brain more finely, into neuron types? For one thing, neurologists can use such information. It's less relevant for stroke or other injuries, which tend to affect all neurons at a particular location in the brain. Some diseases of the brain, however, affect certain types of neurons while sparing others.

Parkinson's disease (PD) starts by impairing the control of movement. Most noticeably, there is a resting tremor, or involuntary quivering of the limbs when the patient is not attempting to move them. As the disease progresses, it can cause intellectual and emotional problems, and even dementia. The cases of Michael J. Fox and Muhammad Ali have raised public awareness of the disease.

Like Alzheimer's disease, PD involves the degeneration and death of neurons. In the earlier stages, the damage is confined to a region known as the basal ganglia. This hodgepodge of structures is buried deep inside the cerebrum, and is also involved in Huntington's disease, Tourette syndrome, and obsessive-compulsive disorder. The region's role in so many diseases suggests that it is very important, even if much smaller than the surrounding cortex.

One part of the basal ganglia, the *substantia nigra pars compacta,* bears the brunt of degeneration in PD. We can zero in even further to a particular neuron type within this region, which secretes the neurotransmitter dopamine, and is progressively destroyed in PD. There is currently no cure, but the symptoms are managed by therapies that compensate for the reduction of dopamine.

Neuron types are important not only in disease but also in the normal operation of the nervous system. For example, the five broad classes of neurons in the retina — photoreceptors, horizontal cells, bipolar cells, amacrine cells, and ganglion cells — specialize in different functions. Photoreceptors sense light striking the retina and convert it

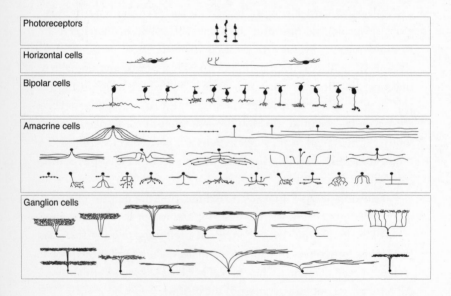

FIGURE 38. *Types of neurons in the retina*

into neural signals. The output of the retina leaves through the axons of ganglion cells, which travel in the optic nerve to the brain.

These five broad classes have been subdivided even further into over fifty types, as shown in Figure 38. Each strip represents a class, and contains the neuron types that belong to the class. The functions of retinal neurons are much simpler than that of the Jennifer Aniston neuron. For example, some spike in response to a light spot on a dark background, or vice versa. Each neuron type studied so far has been shown to possess a distinct function, and the effort to assign functions to all types is ongoing.

As I'll explain in this chapter, it's not as easy as it sounds to divide the brain into regions and neuron types. Our current methods date back over a century to Brodmann and Cajal, and they are looking increasingly outmoded. One major contribution of connectomics will be new and improved methods for carving up the brain. This in turn will help us understand the pathologies that so often plague it, as well as its normal operation.

A modern map of a monkey brain (see Figure 39) brings back pleasant memories of my father's atlas. Its colorings bear mysterious acronyms, and sharp corners punctuate its gentle curves. But maps are not always so charming. Let's not forget that armies have clashed over lines drawn on them. Likewise, neuroanatomists have waged bitter intellectual battles over the boundaries of brain regions.

We already encountered Korbinian Brodmann's map of the cortex. How exactly did he create it? The Golgi stain allowed neuroanatomists to see the branches of neurons clearly. Brodmann used another important stain, invented by the German neuroanatomist Franz Nissl, which spared the branches but made all cell bodies visible in a microscope. The stain reveals that the cortex (Figure 40, right) resembles a layer cake (left). Cell bodies are arranged in parallel layers that run throughout the entire cortical sheet. (The white spaces between cell bodies are filled with entangled neurites, which are not marked by the Nissl stain.) The boundaries in the cortex are not as distinct as those of the

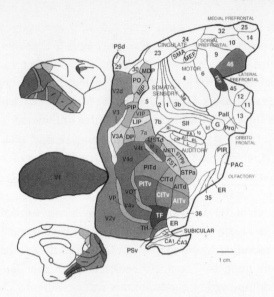

FIGURE 39. *Map of the rhesus monkey cortex laid out flat*

cake, but expert neuroanatomists can make out six layers. This piece
of cortical cake, less than one millimeter wide, was cut from a particular location in the cortical sheet. In general, pieces from different locations have different layerings. Brodmann peered into his microscope
to see these differences, and used them to divide the cortex into forty-
three areas. He claimed that the layering was uniform at every loca-

FIGURE 40. *Layers: cake* (left) *and visual association cortex* (right)

tion within one of his areas, changing only at the boundaries between areas.

Brodmann's map of the cortex may be famous, but it shouldn't be taken as gospel truth. There have been plenty of other contenders. Brodmann's colleagues in Berlin, the husband-and-wife team of Oskar and Cécile Vogt, used a different kind of stain to divide the cortex into two hundred areas. Still other maps were proposed by Alfred Campbell working in Liverpool, Sir Grafton Smith in Cairo, and Constantin von Economo and Georg Koskinas in Vienna. Some borders were recognized by all researchers, but others sparked discord. In a 1951 book Percival Bailey and Gerhardt von Bonin erased most of the borders of their predecessors, leaving just a handful of large regions.

Even worse controversies have plagued Cajal's program to classify neurons into types. He did this based on their appearance, much as a nineteenth-century naturalist would have classified different species of butterflies. One of his favorite neurons was the pyramidal cell. He called it the "psychic cell," not because he believed in the occult, but because he thought that this type of neuron played an important

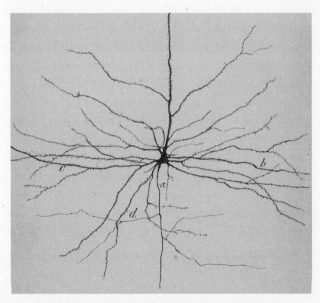

FIGURE 41. *Drawing of a pyramidal neuron by Cajal*

role in the highest functions of the psyche. In Figure 41, a drawing by Cajal himself, you can see the defining features of this neuron type: the roughly pyramidal shape of its cell body, the thornlike spines protruding from its dendrites, and the long axon traveling far away from the cell body. (The axon is directed downward in this image, descending into the brain. The most prominent "apical" dendrite leaves the apex of the pyramid and travels upward, toward the surface of the cortex.)

The pyramidal cell is the most common type of neuron in the cortex. Cajal observed other cortical neurons that had shorter axons, and smooth rather than spiny dendrites. The shapes of nonpyramidal neurons were more diverse, so they were divided into more types, which earned picturesque names such as "double bouquet cell."

Cajal classified neurons into types all over the brain, not just in the cortex. This alternative method of carving up the brain was much more complex than Brodmann's, because every brain region contains many neuron types. Furthermore, the types in each region are intermingled, like different ethnic groups living in the same country. Cajal could not complete the task in his lifetime, and even today this enterprise has only just begun. We still don't know how many types there are, though we know the number is large. The brain is more like a tropical rainforest, which contains hundreds of species, than a coniferous forest with perhaps a single species of pine tree. One expert has estimated that there are hundreds of neuron types in the cortex alone. Neuroscientists continue to argue over their classification.

Their disagreements are a sign of a more fundamental problem: It's not even clear how to properly define the concepts of "brain region" and "neuron type." In Plato's dialogue *Phaedrus,* Socrates recommends "division . . . according to the natural formation, where the joint is, not breaking any part as a bad carver might." This metaphor vividly compares the intellectual challenge of taxonomy to the more visceral activity of cutting poultry into pieces. Anatomists follow Socrates literally, dividing the body by naming its bones, muscles, organs, and so on. Does his advice also make sense for the brain?

"Carving nature at its joints" means cutting in places where connections are weakest. It doesn't take a professional to divide the brain into two hemispheres by cutting the corpus callosum. But most brain regions are not so obvious. The boundaries between cortical areas don't seem like "joints" in the cortical sheet. A great many wires extend across them, connecting neurons on either side.

Of course, we've already carved the brain into extremely fine pieces: individual neurons. No one argues over whether these divisions are objectively defined, now that the debate between Golgi and Cajal has been settled. But as I mentioned in relation to research on Parkinson's disease, it's also useful to divide the brain more coarsely, into regions and neuron types. How can we make these divisions more accurate?

I believe that connectomes will give us new and better ways to divide the brain. We will have to follow Socrates less literally. Unlike poultry, connectomes will be carved in a more abstract way, by classifying neurons based on their connectivity. This approach was previously used to divide the three hundred neurons of *C. elegans* into over one hundred types. The researchers followed a basic principle: If two neurons are connected to similar or analogous partners, they should be grouped in the same type. Some types are trivial, consisting of only a neuron and its twin on the opposite side of the body. The left and right neurons of such a pair are connected to analogous neurons, much as your left arm is connected to your left shoulder and your right arm to your right shoulder. Other types are less trivial, containing up to thirteen neurons with similar connectivity.

Using neuron types, we can simplify the diagram of the *C. elegans* connectome shown in the Introduction (Figure 3). Collapse all neurons of one type into a single node, and repeat this for all types. Figure 42 shows a portion of the result. Each three-letter acronym specifies a neuron type involved in egg-laying behavior. "VCn," for example, stands for the neurons VC1 through VC6, which control the vulval muscles. The lines between the nodes represent connections between neuron types rather than neurons, so we might call the diagram a *neuron type connectome.*

This example shows that carving a connectome not only yields neuron types but tells us how they are connected. Neuroscientists would

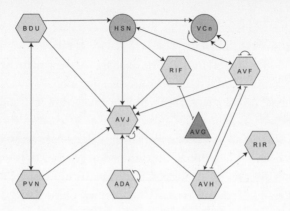

FIGURE 42. *Portion of the "reduced" connectome (with neurons grouped by type) of*
C. elegans

like to do the same for the retina. The connections between the five
broad neuron classes are already known. For example, horizontal cells
receive excitatory synapses from photoreceptors, and send back inhibitory synapses. They also make electrical synapses with each other.
But recall that the five classes are divided into more than fifty neuron
types. Their connectivity is mostly unknown but could be discovered
by finding and carving the neuronal connectome of the retina.

It's worth noting that this approach is different from the classical
one. Cajal first defined neuron types by shape and location, and then
moved on to investigate their connections. Here I'm proposing to turn
this around, starting from connectivity and working backward to define neuron types.

This approach, though different, can still be viewed as a refinement
of Cajal's if we regard shape and location as proxies for connectivity.
To understand why, imagine looking at two neurons. Each neuron extends its branches over some region. If the two regions are completely
separate from each other, for genetic or other reasons, then there is no
way for the neurons to be connected. Contact is a prerequisite for connection, and contact is governed by location and shape.

If shape and location are so closely related to connectivity, why is
connectivity the better approach? The answer is the connectionist
mantra, "The function of a neuron is chiefly defined by its connec-

tions with other neurons." Connections are directly related to function, while shape and location are only indirectly so.

A similar strategy can be applied to divide the brain more coarsely into regions rather than neuron types. In the discussion of rewiring I mentioned that each cortical area possesses a unique "connectional fingerprint," or pattern of connection with other cortical areas as well as regions outside the cortex. We could turn this around and use it to define cortical areas. If we carve a connectome into groups of contiguous neurons, so that each group shares a common connectional fingerprint, we should end up with brain regions. (We'd have to constrain the groups to not overlap in space, or else they might end up being intermingled neuron types rather than spatially distinct regions.)

How is this related to Brodmann's use of layering to define cortical areas? Again, layering should be regarded as a proxy for connectivity. For example, areas 17 and 18 differ by the thickness of layer 4, because of their different connectivity. Layer 4 of area 17 is swollen with many neurons that receive connections from pathways originating in the eyes. The adjacent area 18 doesn't receive such axons, so its layer 4 is not as large.

If layering is so closely related to connectivity, why should we prefer a definition based on the latter? Once again, it's because the layering is less fundamental. The fact that visual pathways lead to area 17 tells us immediately that its function is visual. The fact that area 17 has a thicker layer 4 is only indirectly related to its function.

Brodmann relied on cortical layering and Cajal on neural shape and location. Although more sophisticated than size, these properties were still only crude substitutes for what really matters: connectivity. More than a century later, we should be able to dispense with the proxies and work directly with connectomes.

I've argued that the ideal way of dividing a brain is to carve its connectome. As a bonus, we also learn how the divisions are connected with each other, obtaining a regional or neuron type connectome. How can these simplified versions of the neuronal connectome help us understand the brain?

The importance of regional connections was recognized as early as the nineteenth century, when Wernicke hypothesized a bundle of long axons connecting Broca's and Wernicke's regions. Damage to the bundle, while leaving both speech comprehension and production intact, would make it impossible to repeat words after hearing someone else speak them. Wernicke's region could receive the words but could not relay them to Broca's region to be spoken aloud. Because this hypothetical disorder was due to loss of signal conduction, Wernicke called it *conduction aphasia.* Patients with these symptoms were later discovered, confirming Wernicke's prediction. Furthermore, neuroanatomists identified the hypothetical connection between Broca's and Wernicke's regions as a bundle of axons called the arcuate fasciculus (see Figure 43).

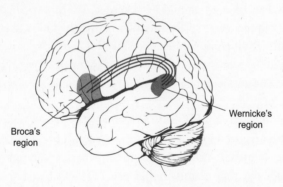

FIGURE 43. *A bundle of axons connecting Broca's and Wernicke's regions*

The Broca–Wernicke model of language shows how one might go about using the regional connectome once it's found. Associate every brain region with an elementary mental function, such as speech comprehension or production. Then explain more complex mental functions, like speech repetition, as combinations of elementary functions. These are carried out by cooperation between multiple regions, which is mediated by regional connections.

Neurologists use this conceptual framework to diagnose patients with brain damage. Damage to a region impairs the corresponding elementary function. Damage to a *connection* impairs complex functions requiring cooperation between regions. Because this paradigm includes connections and allows for distributed functions, it goes beyond localizationism. It is sometimes called connectionism, though it's a different flavor from the neural connectionism introduced earlier. We can also imagine a connectionism based on neuron types. This kind of brain model would be more sophisticated than the neurologists', and far more challenging to construct, because neuron types and their connections are so numerous.

But in the immediate future, a regional connectome seems like the most useful kind for psychologists and neurologists. Olaf Sporns and his colleagues pointed this out when they originally coined the term *connectome* in a 2005 paper. You may have heard of the $30 million Human Connectome Project, which was launched in 2010 by the U.S. National Institutes of Health (NIH). Most people don't realize that this project is only about regional connectomes, and has nothing to do with neuronal connectomes.

I personally spend more time focused on neurons than on regions, but I agree with Sporns and his colleagues about the importance of finding regional connectomes. My only disagreement has to do with the methods. In my view, we need to see neurons to find regional connectomes. In other words, I am still a neuronal chauvinist — but about means rather than ends.

I believe it would be best to find regional connectomes by carving neuronal connectomes, though I admit that this strategy is idealistic at present. In the near term it will be practical only for very small brains, not for human brains. That's why the Human Connectome Project is attempting a shortcut: finding regional connectomes using MRI. As I will explain later, this imaging method is bound to encounter difficulties, because of its limited spatial resolution. In Chapter 12 I will propose an alternative shortcut to finding regional connectomes, one that might be practical in the near future without cutting as many corners. A similar shortcut could also help us find neuron type connectomes.

Not every neuroscientist is convinced that we should put more effort into dividing the brain; some believe that our maps are already good enough. To counter this idea, let's take a closer look at the Broca–Wernicke model of language. It sounds so successful in the textbooks, but the true story is messier.

In Broca's original patient, the brain lesion covered much more than Broca's region, extending into surrounding cortical areas as well as regions below the cortex. It turns out that a lesion of Broca's region alone does not produce Broca's aphasia, and lesions that spare Broca's region can produce Broca's aphasia. The regional basis of Wernicke's aphasia is similarly murky. Furthermore, the double dissociation of speech production and comprehension is not as clean as the textbooks make it sound. For example, Broca's aphasia is usually accompanied by problems with comprehension of sentences. Consistent with these clinical findings, recent fMRI studies have shown that language is less localized than previously thought, involving cortical and subcortical areas beyond those of Broca and Wernicke. Clinical studies do not support the traditional story that conduction aphasia is caused by lesions of the arcuate fasciculus. More embarrassingly, some researchers now deny that the arcuate fasciculus connects Broca's and Wernicke's regions, even though we've believed it does for over a century. Some neuroscientists have found other pathways that do connect these regions.

For all these reasons, language researchers are struggling to formulate a replacement for the Broca–Wernicke model. The new model will have to include additional cortical areas, as well as brain regions outside the cortex, and will have to explain a more complex array of linguistic abilities than the simplistic duo of speech comprehension and production. Everyone agrees on the need for an improved model, but there is no consensus on how to find it. I don't presume to know, but I am sure that better maps of the brain would be helpful.

Dividing the brain has historically been more art than science. Like a physician's divining a disease from a constellation of symptoms or a judge's compromise between multiple legal precedents, dividing the brain has never been reduced to a simple formula. Some of the boundaries between brain regions are no doubt arbitrary, the result of historical accidents and errors of neuroanatomists. Like globes and atlases, our maps of the brain do not represent objective, timeless truth. Sometimes new regions are created, or the borders between them may shift. Disputes over borders can erupt in acrimonious debate between scientists, ideally settled peacefully by the patient negotiations of committees.

We should not become complacent with this state of affairs. Our current maps of the brain may not be as poor as world maps from centuries ago, which look almost laughable to our modern eyes, but there is plenty of room for improvement. Maps by themselves will not tell us how brain regions contribute to mental function. They will, however, accelerate research by providing a firm foundation on which to stand.

My emphasis on structural criteria for dividing the brain will seem strange to a contemporary neuroscientist, who is used to combining them with functional criteria. But this kind of emphasis is commonplace in the rest of biology. The organs of the body were known as structural units long before their roles were understood, and can be identified by a naive observer with no knowledge of function. Similarly, the organelles of a cell were observed in the microscope long before it was known that the nucleus contains the genetic information and that the Golgi apparatus packages proteins and other biomolecules before sending them to their proper destination.

In general, biological units are both structural and functional entities, but they are typically identified first by structure; their functions are figured out later. Brain regions and neuron types should be the same way. Neuroscientists following Brodmann and Cajal have long pursued the structural approach to dividing the brain but have achieved only partial success. The problem is not that the approach is fundamentally flawed; rather, it is that our techniques for measuring brain structure have been inadequate. Any division of the brain is only as good as the data on which it is based. By providing dramatically better data about structure, connectomics will allow a more objective division of the brain, and by extension the mind.

Identifying cortical areas by examining the symptoms of brain damage is analogous to the way that the Austrian monk Gregor Mendel identified genes in the 1860s. His experiments on the interbreeding of plants showed that the inheritance of certain traits (now called Mendelian) was controlled by variations in a single unit, later called a gene. In his simple picture, traits and genes had a one-to-one correspondence. But now we know that most traits are not Mendelian. Most traits can be affected by many genes, and one gene may affect many traits. This is because a gene encodes for a protein, which can perform many tasks.

Similarly, localizationism attempted to establish a one-to-one correspondence between mental functions and cortical areas. But it turns out that most mental functions require the cooperation of multiple cortical areas, and most cortical areas participate in multiple mental functions. This makes it problematic to use functional criteria to define cortical areas. The right strategy is to identify the areas by structural criteria and then understand how the interactions between areas give rise to mental functions. This approach will become practical as our technologies improve.

We expect to find the same regions and neuron types in all normal brains. The regional and neuron type connectomes are likely to vary little across normal individuals, and are likely to be highly determined by genes. As I mentioned earlier, genes guide the growth of the branches of neurons, thereby influencing the neuron type connectome. Scientists are also identifying genes that control the formation of cortical areas. Your mind and mine may be similar because our brain regions and neuron types are connected in the same ways.

In contrast, neuronal connectomes will vary greatly across individuals and will be strongly influenced by experiences. These are the connectomes we must study if we want to understand human uniqueness. And we should examine them for traces of the past, for what could be more integral to our uniqueness than our own memories?

ELEVEN

Codebreaking

I LIKENED THE TASK of seeing connectomes to navigating the twisting passages of the Labyrinth. According to legend, this structure was located near the palace of King Minos at Knossos on the island of Crete. In 1900 Knossos yielded a second metaphor for the brain. Hundreds of clay tablets were excavated from the ancient ruins. Their discoverer, the British archaeologist Arthur Evans, could not read them, for they were inscribed in an unknown language. For decades the tablets remained unintelligible, and their mysterious script became known as Linear B. Finally, in the 1950s, Michael Ventris and John Chadwick succeeded in decoding Linear B, and the tablets' meaning was revealed.

Once we can see connectomes and carve them up into pieces, the next challenge will be to decode them. Will we learn to understand their language? Or will their patterns of connectivity merely tantalize, refusing to give up their secrets? Decoding Linear B took half a century, but at least Ventris and Chadwick succeeded in the end. Similar attempts have failed for a number of lost languages. Linear A, the script used in ancient Crete before Linear B, remains incomprehensible. The Indus script of ancient Pakistan, the Zapotec writing system of ancient Mexico, and the Rongorongo glyphs of Easter Island have also eluded decipherment.

What exactly does it mean to decode connectomes? Sometimes it's easier to understand a concept if you consider its most extreme version. For a thought experiment, let's imagine that you are living in the far future. Medicine has become very advanced, but, alas, your great-great-grandmother finally dies (at age 213). You take her to a facility that slices up her brain, images the slices, finds her connectome, and hands you a little electronic stick containing the data. After you get home, you are feeling sad because you miss talking to her. (She was your favorite great-great-grandmother.) You place the stick in your computer and ask it to recall some of her memories, and soon you are feeling better.

Will it ever be possible to read memories from connectomes? I proposed a similar thought experiment earlier, asking you to consider whether someone could read your perceptions and thoughts by measuring and decoding the spiking of every neuron in your brain. Some neuroscientists believe we could do this if our technologies for measuring spiking were advanced enough. Why do they think so? From the spiking of a Jennifer Aniston neuron we can already guess whether a person is perceiving Jen. Neuroscientists extrapolate from this small success that the spiking of all neurons would give us a complete picture of our thoughts and perceptions.

Similarly, we might believe that memories could be read out from connectomes, if only there were a small success in this direction. Finding an entire human connectome still lies far in the future. For now, we have to work with partial connectomes derived from small pieces of brain. Perhaps we could pick a small chunk of human brain and attempt to read out memories from it. Or a piece of animal brain?

One thing is certain: Seeing the connectome will be just the first step. To read a book, you must do more than see the text. You must know the language in which the book is written, not to mention the letters of the alphabet and the spellings of words. In more technical terms, you must know how information is *encoded* in the marks on paper. Without knowledge of the code, a book is just a bunch of meaningless markings. Likewise, to read memories, we will have to do more than see connectomes. We will have to learn how to decode the information they contain.

In what region of the human brain are we likely to find memories? Important clues come from the life of Henry Gustav Molaison, who died at a nursing home in Connecticut in 2008. During his lifetime he was known to the world as H.M., to protect his privacy. Many doctors and scientists studied H.M., who became one of the most famous neuropsychological cases since Broca's patient Tan.

In 1953, at the age of twenty-seven, H.M. underwent surgical treatment for severe epilepsy. The surgeon believed that H.M.'s seizures were originating in the medial temporal lobe (MTL), so he removed this region from both sides of H.M.'s brain. After surgery, H.M. appeared normal. His personality, intellect, motor skills, and sense of humor were intact. But there was an important and utterly incapacitating change. For the rest of his life, H.M. woke up every morning in a hospital room with no idea why he was there. He could not learn the names of the caretakers he saw every day. He could not name the president or describe current events. In contrast, H.M. could still remember events from his life that had occurred before the surgery. The MTL seemed essential for storing new memories but not for retaining old ones.

You probably remember that Itzhak Fried and his colleagues found the Jennifer Aniston and Halle Berry neurons in the MTL, showing that this region is involved in both perception and thought. Further experiments have explored its role in memory recall. A patient was first shown many short video clips (five to ten seconds each) from cartoons, sitcoms, movies, and so on, while the activity of an MTL neuron was recorded. Later on, the patient was asked to freely recall the videos and report verbally whenever one came to mind. (During this second part of the experiment, no video was shown to the patient.)

One neuron spiked when the patient viewed a video of Tom Cruise but was much less active for other celebrities, places, and so on. Later on, the same neuron spiked whenever the patient reported recalling Tom Cruise but not when recalling other videos. Other neurons behaved similarly, being selectively activated by viewing or recalling one video but not others.

Perhaps the Tom Cruise neuron belongs to a cell assembly in the MTL. Perceiving or recalling Tom Cruise activates the cell assembly and hence the Tom Cruise neuron. If we want to read memories from connectomes, why don't we start by looking for cell assemblies in MTL? Unfortunately, the MTL is so large a region that finding its connectome is too difficult to be practical with our current technology.

We could narrow our search by focusing on the hippocampus, a part of the MTL thought to be important for storing new memories. In particular, the CA3 region of the hippocampus contains neurons that make synapses onto each other. Perhaps these connections allow groups of CA3 neurons to form cell assemblies. But the human CA3 region is still rather large, so finding its connectome is currently out of reach. If we want to read memories, we had better find a smaller piece of brain to start with.

H.M.'s amnesia only applied to *declarative* memory, which involves information that can be explicitly stated or "declared." It includes autobiographical events ("I broke my leg skiing last year") as well as facts about the world ("Snow is white"). This is the most common meaning of the term *memory*.

There are also *nondeclarative* forms of memory, which involve information that is implicit rather than explicitly stated. These include motor skills and habits. H.M. could learn new motor skills, such as tracing a shape with a pencil while viewing his hand in a mirror. Based on his case and other types of evidence, neuroscientists have concluded that declarative and nondeclarative memory are distinct faculties, and perhaps served by different brain regions.

These two types of memory share some features, however. In his treatise *On Memory*, Aristotle compared recollection to movement: "Acts of recollection, as they occur in experience, are due to the fact that one movement has by nature another that succeeds it in regular order." One can imagine that sequential memories, whether declarative or nondeclarative, are retained in the brain as synaptic chains. Perhaps the finger movements of a piano sonata played from memory are driven by sequential spiking of a synaptic chain somewhere in the pianist's brain.

It's difficult to study declarative memory in animals, who can't tell us what they recall. But animals are perfectly capable of storing implicit memories. Why not try to read these from animal connectomes? I propose that we do this by searching for synaptic chains in the brains of birds.

Although birds are warm-blooded animals like us, they are more distant evolutionary cousins of ours than rodents. Since they don't nurse their young, they're not classified as mammals. But mammals have no monopoly on intelligence. Despite the use of "birdbrain" as an insult, birds are actually quite smart. Mockingbirds and parrots excel at vocal mimicry, and crows can count and use tools. Because of these sophisticated behaviors, neuroscientists have been increasingly interested in our avian relatives.

Many study the zebra finch, a small Australian native that has spread throughout the world as a lovely pet. Males are adorned with orange cheeks and striking black and white patterns over the rest of their bodies. The male finch in Figure 44 is singing to the female, inviting her to mate. Males of other species also sing to warn other males to keep out of their territory. All that twittering isn't meant for us, but it sounds beautiful just the same. Other birds too are popular pets because of their singing ability—canaries, for instance. Mozart kept a pet starling, and taught it to trill a theme from the finale of one of his concertos. (Some claim the opposite, that the bird inspired his compositions.) Since birdsong makes use of pitch, rhythm, and repetition, some call it "nature's music." Others compare it to language, as when the nineteenth-century master Percy Bysshe Shelley wrote of his art, "A poet is a nightingale who sits in darkness and sings to cheer its own solitude with sweet sounds."

FIGURE 44. *Male zebra finch singing to female*

You might think that song is instinctive. Perhaps baby birds spring from the egg already knowing how to sing? No; those who have suffered through piano lessons need not be envious. The zebra finch does not acquire its talents effortlessly. Before starting to make sounds, a young male first hears his father's song. Later he starts to "babble," like a human baby making nonsense sounds. Over the next few months he practices singing tens of thousands of times, and ultimately learns to copy his father's song.

As a mature adult, a zebra finch sings essentially the same song every time. He does not improvise like a jazz pianist; he's more like a skater tracing compulsory figures on the ice. The song is said to be "crystallized." The bird has stored a memory of its song and can recall it at will.

To produce sounds, birds use a vocal organ called the syrinx, which is like our larynx. Forcing air through the syrinx causes its walls to vibrate like a wind instrument. The pitch and other properties of the resulting sound are controlled by muscles around the syrinx, which in turn receive instructions from the bird's brain. In the 1970s, Fernando

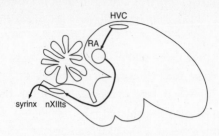

FIGURE 45. *Song-producing regions of the bird's brain*

Nottebohm identified the relevant regions of the brain, shown in the diagram in Figure 45. The names of the regions are long and complicated, so scientists simply use the abbreviations HVC, RA, and nXII.

To understand the roles of the regions, let's compare this system to an artificial one for producing music. Perhaps you have a friend who is fanatic about high-end stereo equipment. Such audiophiles aren't satisfied with all-in-one systems; they like to have many separate components. In your friend's expensive stereo system, the compact disc player generates electrical signals, which travel to the preamplifier and then to the amplifier, and are finally transformed into sounds by the loudspeakers. In a bird's brain, electrical signals travel along an analogous pathway from HVC to RA to nXII, and are finally converted into sounds by the syrinx. Every time the stereo plays Beethoven's Fifth Symphony, both the electrical signals in its components and the sounds from the loudspeaker are repeated in exactly the same sequence. Likewise, both the sounds from the syrinx and the spikes of neurons are repeated exactly the same way every time the bird sings.

Let's take a closer look at HVC. This region comes first in the song pathway, like the compact disc player of the stereo. Its name was originally "hyperstriatum ventrale, pars caudale," or "HVc" for short. Later on, Nottebohm changed the name to "high vocal center," abbreviated "HVC." In 2005 a committee of neuroscientists decided that the letters stand for nothing. (The situation is like that of the SAT, which once stood for "Scholastic Aptitude Test" and then for "Scholastic Assessment Test"; now its owner and developer, the College Board, offers no meaning at all.)

The name change came about because Harvey Karten, a specialist in brain structure and evolution, had convinced his colleagues that birds' brains are more similar to ours than previously thought. Neuroscientists had previously considered HVC analogous to the mammalian striatum, which is part of the basal ganglia, and believed that birds lack anything comparable to the neocortex. But Karten argued that a region called the dorsal ventricular ridge functions like the neocortex. It contains a number of subregions that are believed important for the sophisticated bird behaviors mentioned above. One of these subregions is HVC.

Michale Fee and his collaborators have measured spikes in HVC in live birds while they're singing. Some HVC neurons send their axons to RA, and they're the ones of interest here, as their signals travel along the song pathway. A zebra finch song consists of a few repetitions of a single motif. During a motif, which lasts 0.5 to 1 second, the neurons spike in a highly stereotyped sequence. In Figure 46 I've cartooned the spikes of three neurons. Each neuron waits until its moment in the motif comes, spikes for a few milliseconds, and then falls silent again. The time of spiking is precisely locked to a particular moment of the motif. This kind of sequential spiking is exactly what we'd expect from a synaptic chain.

As Beethoven booms from the stereo system, the loudspeaker vibrates while electrical signals in the stereo fluctuate wildly. Unlike the

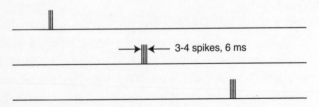

FIGURE 46. *Cartoon depicting the spikes of three neurons in area HVC of the zebra finch brain*

fleeting signals, the compact disc remains serene and unchanging. Underneath its label, the plastic surface contains hundreds of millions of microscopic indentations, which encode music as bits of digital information. The plastic will maintain its shape for decades, as the manufacturer guarantees; that stability is why the compact disc will reproduce Beethoven over and over again. Its *material structure* enables it to retain a "memory" of Beethoven's music.

I've compared the spikes of HVC neurons to electrical signals in your compact disc player. Now I'd like to take the analogy further and propose that the HVC connectome is like the compact disc. Let's suppose that it contains a synaptic chain, which no longer changes once a song has crystallized in the adult male. According to this proposal, the HVC connectome retains the memory of the song. Whenever the bird sings, the memory is recalled by converting it into sequential spiking. These signals are fleeting; the material structure of the connections in HVC, however, remains unchanged.

HVC is just a fraction of a cubic millimeter in volume. It should be technically feasible to find its connectome in the near future. Then we could simply examine the connectome to find out whether it's organized like a synaptic chain. This will require some analysis, because it's not obvious whether a connectome contains a chain unless the sequential ordering of the neurons is known. To see why, consider the diagrams shown in Figure 47, both of which have exactly the same connectivity. The neurons on the left have been scrambled to hide the chain. To reveal it, we must unscramble the neurons to yield the diagram on the right. You can try doing this by hand for our small made-up connectome. But a real HVC connectome is complex enough that a computer would be necessary.

Suppose we succeeded in unscrambling the HVC connectome. From the resulting chain we'd be able to guess the order in which the neurons spike during the song. This would amount to reading the memory of song, in the sense that we could guess the activity sequence that is replayed in HVC when the bird sings.

How can we confirm that our reading is correct? Ventris and Chadwick convinced the world they had decoded Linear B because their

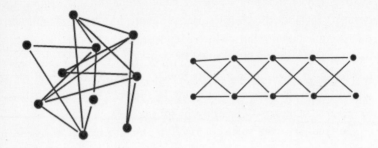

FIGURE 47. *Synaptic chains, scrambled* (left) *and unscrambled* (right)

reading of the clay tablets made sense. If they had failed, the deciphered text would presumably have been gibberish. A stronger test than internal consistency would be to observe and speak with the people who wrote the tablets, but the impossibility of time travel prevents us from doing so.

Similarly, if unscrambling the HVC connectome revealed a synaptic chain, we could already be confident about our reading. Unlike Ventris and Chadwick, we could obtain more conclusive proof without resorting to time travel. Suppose another neuroscientist measures the spike times of HVC neurons during singing but withholds them in order to quiz us. We find the HVC connectome and then read it to guess the spike times. Our examiner compares our guesses with the real spike times; if they match, our reading of the connectome is correct.

To measure the spike times of HVC neurons, our examiner could obtain help from chemists, who have invented ways of staining neurons so that they resemble blinking lights in the microscope, glowing when they spike and darkening when they fall silent. The images from our examiner's light microscope would also tell us the exact locations of the cell bodies of the HVC neurons. Later on, these locations could be matched up with cell bodies in electron microscopic images of the dead brain. Establishing this correspondence would enable our examiner to compare the real spike times of HVC neurons with the times guessed by reading the connectome.

Of course, it's always possible that we will fail to unscramble the

HVC connectome. We might not be able to order the neurons so that the synapses respect the sequential rule of connection. In other words, no matter how we arrange the neurons in a sequence, there are many connections that go backward or jump too far ahead. This would mean that the HVC connectome is not organized like a synaptic chain. Such failure would still be progress. For the purpose of advancing science, rejecting models is as important as confirming them.

If the HVC connectome does turn out to be organized like a chain, that would be evidence that it helps retain the bird's memory of its song. But how do memories like this get stored in the first place? Some theoretical neuroscientists have proposed that HVC neurons in young males are initially driven by random input from some other source. This activates the neurons in random sequences, some of which become reinforced by Hebbian strengthening of connections. These select sequences start to occur more often, thus becoming further reinforced. Ultimately a single sequence is reinforced so much that it dominates all others. This sequence corresponds to the final synaptic chain that we suspect exists in adult males.

According to this proposal, reweighting stores the memory of song. Synapses change in strength, but they are not created or eliminated. The unweighted connectome, which omits information about synaptic strengths, does not contain any of the information in the memory. There is no chance of reading out the spike times of neurons from it. Only the weighted connectome is readable, as only the strong synapses are organized in a chain. In other words, connectomes must include synaptic strengths if they are to be decoded. In principle, that's no problem for connectomics. It should be possible to estimate the strengths of synapses from their appearance in electron microscopic images. As I mentioned earlier, synapses are thought to grow bigger when they get stronger, so size is correlated with strength. Future research should be able to tell us the accuracy of this method for estimating synaptic strength.

Another possibility is that reconnection also plays a role in the storage of song memory. Maybe synapses not involved in the synaptic chain weaken as the bird learns, and are eventually eliminated. If reconnection does play a role, then even the unweighted connectome might be readable. By attempting to read both unweighted and weighted versions of the HVC connectome, we could conceivably distinguish between the pure reweighting theory of memory and the reweighting-plus-reconnection theory.

Neuroscientists have hypothesized that the other two R's of connectome change — rewiring and regeneration — also play a role in memory storage, but there is little empirical evidence one way or the other. Fernando Nottebohm and his collaborators have studied regeneration in the brains of canaries and other songbirds. They have shown that HVC shrinks because neurons are eliminated during the part of the year when canaries don't sing. When the singing season comes around again, HVC expands by creating new neurons. Nottebohm's research on regeneration played an important historical role in reviving neuroscientists' interest in the subject, but the function of regeneration remains unclear.

This question could be investigated in a number of interesting ways if the synaptic chain model of HVC is correct. During the off season, does a dormant synaptic chain continue to store the memory of song? When the new neurons enter HVC, do they integrate into the chain? If so, how do they do it? Neural Darwinism predicts that newly created neurons are randomly connected with other neurons. This prediction could be tested empirically by connectomics, with the aid of special stains that mark new neurons.

Similar questions can be asked about the elimination of neurons. What causes neurons to commit suicide? Is it triggered by elimination of synapses and branches, which in turn happens because the neuron fails to integrate into the chain? This hypothesis could be probed using connectomics, through snapshots of neurons caught during the process of dying. To prepare for the off season, are neurons eliminated in such a way that prevents the chain from breaking?

Because of technical limitations, neuroscientists have had to settle for counting increases and decreases in the number of neurons. These studies suggest that regeneration is important, but they do not reveal its exact role in memory. To make further progress, it's crucial to know how new neurons get wired up to the existing organization, and whether the elimination of neurons depends on how they are wired. This kind of information can be provided by connectomics. The function of rewiring could also be studied in HVC by investigating how the growth and retraction of branches of neurons depends on their connections with other neurons.

I've outlined a plan for finding synaptic chains in the HVC connectome and cell assemblies in the CA3 connectome. I've called this "reading memories" from a connectome. More precisely, I've proposed a way of analyzing connectomes to guess activity patterns that are replayed during recollection of a memory. But let me emphasize: That's not the same as knowing what the memory *means*. By analyzing the HVC or CA3 connectomes, we won't know what the bird's song sounds like, or what's in the videos that were previously seen by a human research subject. We might call this the reading of an "ungrounded" memory, one that is divorced from its meaning in the real world.

I already proposed one way of grounding the memory, which is to measure HVC activity in birds as they sing, or CA3 activity in humans while they describe what they're experiencing. Then each neuron could be placed in correspondence with a particular movement or reported idea. This sort of approach uses measurements of spiking in a live brain to ground memories read out after the brain is dead. It's the only approach possible in the near future, as long as we can find only partial connectomes from small chunks of brain.

In the long term, though, I expect that we will be able to find connectomes of entire dead brains. Then it may become possible to ground memories without measuring spiking in live brains. To do this, we'd have to figure out, for example, whether a CA3 neuron is selectively activated by Jennifer Aniston or some other stimulus. Could this be possible by analyzing the pathways that bring information from the sense organs to the CA3 neuron?

It might be, if we employ the hypothesized rules of connection for perceptual neurons — for example, "A neuron that detects a whole receives excitatory synapses from neurons that detect its parts." The Jennifer Aniston neuron might receive inputs from a "blue-eye neuron," a "blond-hair neuron," and so on.

For now, researchers are starting to test this part–whole rule by combining measurements of spiking with connectomics in animals. The first step is to determine the functions of neurons in perception by measuring their spiking in response to various kinds of stimuli, as in the Jennifer Aniston experiment. This is done as described earlier, by staining the neurons so that they blink when active, and observing the neurons through a light microscope. Then researchers image this particular chunk of the brain using an electron microscope to discover how the neurons are connected. Kevin Briggman and Moritz Helmstaedter have accomplished this feat with retinal neurons, working with Winfried Denk. Studies of neurons in the primary visual cortex have been performed by Davi Bock, Clay Reid, and their collaborators. This approach, as it develops, will make it possible to see whether there are in fact connections between neurons that detect parts and wholes.

In the coming years the part–whole rule of connection will be tested in this way. For the sake of discussion, let's suppose that the rule is true, and speculate about how we could use it to read connectomes. The driving idea behind the rule is that a neuron stands on the shoulders of other neurons. We could start by applying the rule to the neurons near the bottom of the hierarchy and guess which stimuli they used to detect. These are the neurons just one step away from the sense organs. Then we could move step by step up the hierarchy, each time guessing the stimuli that neurons detect from the part–whole rule. Eventually we might reach the top of the heap — CA3 neurons — and guess which stimuli used to activate them in the live brain. (A neuron that receives connections from neurons that detect floppy ears, sad brown eyes, wagging tail, and loud bark — that's the neuron that detected your great-great-grandma's dog.)

Reading memories from dead human brains might sound cool — you could certainly imagine an entertaining movie being built around this plot device — but it's too far off to be considered seriously as an important practical application of connectomics. What I'm proposing instead as a basic research challenge is to decode the HVC connectome. It would be a way of improving our understanding of how the brain's function depends on the connections between its neurons.

I've discussed several ways of analyzing connectomes: carving them into brain regions, carving them into neuron types, and reading memories from them. These approaches may seem quite different, but all can actually be viewed as the formulation of rules of connection governing neurons. Each approach in the list is progressively more accurate at predicting connections, because its rules are based on more specific neuronal properties.

For example, carving the avian brain into regions would yield coarse rules, such as "If two neurons are in HVC, they are likely to be connected to each other." It's certainly true that a connection between two HVC neurons is more likely than a connection between an HVC neuron and, say, a neuron in a visual region called the Wulst, which doesn't happen at all. Nonetheless, this rule would still be lousy at predicting whether two arbitrary HVC neurons are connected, as this turns out to be quite improbable too.

To make the rule more accurate, it might help to divide HVC into multiple neuron types. I didn't mention it before, but our previous discussion was actually specific to just one type of HVC neuron, the one that sends axons ("projects") to RA. This neuron type is of special interest because it generates the kind of sequential spiking characteristic of a synaptic chain. We could use it to formulate a revised rule: "If two HVC neurons both project to RA, they are likely to be connected to each other." This more specific rule could well be more accurate.

Even better would be to make the rule depend on the spike times of the neurons during song: "If two HVC neurons both project to RA, and their spike times during song are one after the other, they are likely to be connected to each other." If the synaptic chain model is correct, then this rule would be highly accurate at predicting connections.

If we really want to understand how the brain works, we need this third kind of rule, which depends on functional properties of neurons as determined by measurements of spiking. The coarser rules of connection, which depend on region or neuron type, get us only part of the way there. Knowing the regional connections that lead from HVC to the syrinx tells us why HVC neurons have functions related to song. But that's not enough for elucidating why different HVC neurons spike at different times during song.

Likewise, knowing regional rules of connection might tell us why the Jennifer Aniston and Halle Berry neurons do similar things — both are activated by visual stimulation — but no fan would say that they do *exactly* the same thing. We'd like to know why the Jennifer Aniston neuron responds specifically to Jen and not Halle, and vice versa. For this we need something like the part–whole rule of connection, which again depends on the functional properties of neurons.

In the most general sense, decoding connectomes means reading out the roles played by neurons not only in memories but also in thoughts, feelings, and perceptions. If we can succeed at decoding, we'll know that we've finally found rules of connection precise enough for understanding how the brain works. And then we'll be ready to return to the question that we started with, the one that motivates this book: Why do brains work differently?

TWELVE

Comparing

N ELEMENTARY SCHOOL my friends and I tried not to gawk at identical-twin classmates, but we couldn't help staring as we strained to tell them apart. Photos of Siamese twins were even more riveting. We looked at them long and hard while flipping through beat-up copies of the *Guinness Book of World Records*. Twins just seemed spooky, though we weren't sure exactly why.

Native American and African myths are full of stories about twins. The Navajo people trace their ancestry to the goddess Changing Woman. Impregnated by sunbeams, she bore twin sons named Monster Slayer and Born for Water. They grew up in twelve days, traveled to find their father, the Sun, and went on to engage in deadly combat with giants and monsters.

Many more twins figure in the world's legends and literature. Fraternal twins have always seemed special, and identical twins perhaps even magical. Why do we feel that way? For one thing, identical twins assault our bedrock assumption that every human being is unique; we're unsettled by their alikeness. But we're also fascinated by the slight differences that are visible if we look closely.

In Greek myths, twins were often the offspring of one mother but two different fathers, one divine and the other mortal, which explained the twins' different natures and fates. Today we know that we can account for those differences by pointing to the genomes of fraternal twins, who share only half of their genes. Identical twins, however, look almost indistinguishable from each other because of their duplicate genomes. I mentioned this claim about identical twins earlier when discussing the genetics of autism and schizophrenia, but it needs some qualification. Recent genomic studies have demonstrated that tiny deviations in DNA sequence arise during the process of twinning, the divergence of a fertilized egg into two embryos. These deviations might explain why identical twins look slightly different, and perhaps why they don't think and act in exactly the same way. But genes do not fully explain mental aspects that depend on learning. Even for twins who remain conjoined (the term that has replaced *Siamese*) instead of being surgically separated, life experiences do not match exactly. Such twins are literally inseparable, but their memories are not identical.

According to connectionist thinking, identical twins have different memories and minds chiefly because their connectomes differ. Many people have wondered what it would be like to have a twin sibling. Sometimes I fantasize about a mad scientist creating my "connectome twin," a person with a brain that is wired exactly like mine. Would I be enraptured to meet him? Would my girlfriend grow jealous of our close relationship, complaining about yet another proof of my narcissistic tendencies? I suppose I could confide anything to my twin, who would be guaranteed to understand me. Then again, maybe it would be boring to pour out my problems to someone who thinks in exactly the same ways.

And what if, after a week of getting to know each other, we were kidnapped by a team of crazed gunmen? Let's say they decide to shoot one of us and send the body along with the ransom note, as proof of abduction. Should I fear being shot, or should I be altruistic and volunteer to take the bullet? Maybe it doesn't matter, as all my memories and personality will survive in my twin even if I die, and vice versa. But wait. A week has passed since the mad scientist breathed life into my replica. Our connectomes have been changing since then. They diverged from the first instant after duplication, so our minds are no longer identical.

Luckily I'll never be forced to engage in the head-scratching required to solve this distressing philosophical dilemma. We won't be seeing human connectome twins any time soon. But what about worms? I referred in the Introduction to "the" connectome of *C. elegans*, implying that any two worms are connectome twins. But is this really true? Certainly the neurons are identical, so we should be able to take two connectomes, match up their neurons one to one, and check to see whether the connections are the same.

Such a comparison has never been done in its entirety, because it would require two complete *C. elegans* connectomes, and finding just one was difficult enough. David Hall and Richard Russell took the shortcut of comparing partial connectomes from the tail ends of worms. They didn't find a perfect match. If two neurons were connected by many synapses in one worm, in all likelihood they were also linked in another worm. But if two neurons were connected by a single synapse in one worm, there might be no synapse at all between them in another.

What caused these variations? The worms had been highly inbred in the laboratory for many generations, by exaggerating the methods used to create purebred dogs and horses. That made all lab worms genomic twins, but a few differences did remain in their DNA sequences. Could these differences account for connectome variation? Or is such variation a sign that worms learn from experience? Or perhaps the variation is due neither to genes nor to experience, but rather to random sloppiness as the worm's neurons wire together during development. Any of these explanations could be true, but more research is needed to test them.

Did connectome variation affect behavior, giving worms distinctive "personalities"? Hall and Russell did not study this question, so we don't know. Their worms were inbred but otherwise normal. Other researchers have identified genetically defective worms that also behaved abnormally. Finding their connectomes has yet to be done, but after that is accomplished, it should be straightforward to compare the connectomes of abnormal and normal worms if the neurons can be placed in one-to-one correspondence. If there are missing neurons, or additional neurons, then matching the connectomes will be a bit more difficult; still, it should be possible. Research of this type will take off as it becomes easier to find *C. elegans* connectomes.

Comparing the connectomes of animals with big brains will be much more challenging. As I mentioned in the Introduction, big brains vary greatly in number of neurons, so there's no way of placing neurons in one-to-one correspondence. Ideally, we would find some way to match up neurons with similar or analogous connectivity. According to the connectionist mantra, such neurons would also have similar functions, like a Jennifer Aniston neuron in one brain and a Jennifer Aniston neuron in another. The correspondence would not be one to one, as the number of Jennifer Aniston neurons might vary across individuals. (Some people might even lack Jennifer Aniston neurons altogether, having never had the benefit of exposure to her.) This kind of matching would require sophisticated computational methods yet to be developed.

An alternative approach is to compare connectomes after coarsening them. We could define reduced connectomes for brain regions or neuron types, as described earlier. Since these are expected to exist in all normal individuals, it should always be possible to place them in one-to-one correspondence. Comparing reduced connectomes of big brains would be as simple as comparing worm connectomes.

Previously I argued that regional or neuron type connectomes would be insufficient for understanding our memories, the most unique aspect of our personal identities. But other distinguishing mental characteristics, such as personality, mathematical ability, and autism, seem more generic than autobiographical memories. These properties of minds might be encoded in reduced connectomes.

• • •

In principle, we could find reduced connectomes by carving up neuronal connectomes. Even for rodent brains, however, finding an entire neuronal connectome is a long way off. An alternative is to develop shortcut methods that find reduced connectomes directly, without requiring neuronal connectomes. Such methods would be technically easier, as they would not require collecting so much image data.

Some neuroscientists would like to use light microscopy to find connectomes for neuron types—an approach pioneered by Cajal, who concluded that two neuron types were connected when one type extended axons into a region occupied by dendrites of the other type. His approach was piecemeal, but with modern technologies it could be applied systematically. To find a neuron type connectome, though, we would have to combine neurons imaged in many brains, as light microscopy can reveal only a small fraction of a single brain's neurons. Therefore, this approach might be less useful for finding differences between individual brains.

Light microscopy could also be used to map regional connectomes. To apply this approach to the cortex, we must map a specific part of the cerebrum that I haven't discussed yet—the cerebral white matter. Recall that the cerebrum atop the brainstem resembles a fruit on a stalk. The "peel" of the fruit is the cortex, otherwise known as the gray matter. Cutting the fruit open reveals its "flesh," called the white matter, as shown in Figure 48.

The distinction between gray and white matter was known in antiquity, but their fundamental difference became clear only after the discovery of neurons. The outer gray matter is a mixture of all parts of neurons—cell bodies, dendrites, axons, and synapses—while the white matter contains only axons. In other words, the inner white matter is all "wires."

Most white-matter axons come from neurons in the surrounding cerebral cortex. They belong to pyramidal neurons, which constitute about 80 percent of all cortical neurons. Earlier I mentioned that this neuron type has a cell body with a triangular or pyramidal shape, and an axon that travels a long distance from the cell body. Let's refine the picture here. The apex of the pyramid points toward the exterior of the

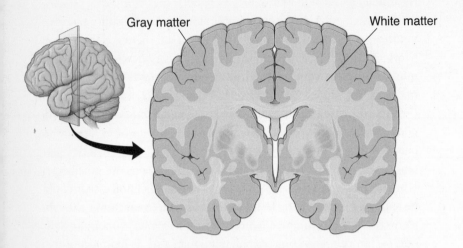

Gray matter
White matter

FIGURE 48. *Gray versus white matter of the cerebrum*

brain. The axon comes straight out of the base of the pyramid, perpendicular to the cortical sheet, and plunges into the white matter, as Figure 49 shows.

As the axon dives down, it sends out side branches, called "collaterals," which are for making synapses onto nearby neurons. But the main branch of the axon finally leaves the gray matter and enters the white matter to start its journey to other regions. In each of its destination

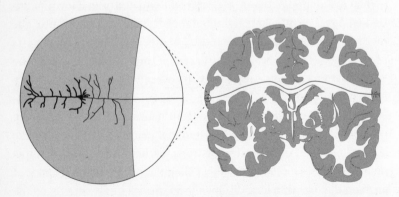

FIGURE 49. *Collateral and main branches of a pyramidal neuron's axon*

regions, it forks out many branches to make connections with neurons there.

Some axons don't travel very far, reentering the gray matter close to where they started. But most axons of pyramidal neurons project to other regions in the cortex, some going as far as the other side of the brain. Some white-matter axons — a small minority — connect the cortex with other structures in the brain, such as the cerebellum, the brainstem, or even the spinal cord. These axons make up less than one-tenth of the white matter. The cortex is highly self-centered, primarily "talking" with itself rather than the outside world.

Here's another way to think about it: If the axons and dendrites in the gray matter are like local streets, the axons of the white matter are like the superhighways of the brain. They are relatively wide and un-branched, and also extremely long. In fact, the total length of these axons is roughly 150,000 kilometers, over a third of the distance from the Earth to the Moon. And herein lies the challenge: Finding the regional connectome requires tracing the journey of every axon in the white matter.

It seems like an impossible task, but it could be done by slicing and imaging all of the white matter and using computers to follow the path traveled by each axon in the images. The start and end points of every path would define a connection between two locations in the cortex. Is this approach too difficult to be practical? After all, the cerebral white matter is comparable in volume to the gray matter, and we are still struggling to reconstruct one cubic millimeter of that. Given this, it might seem outlandish to propose reconstructing hundreds of cubic centimeters of white matter. My proposal seems less crazy once you know that white-matter axons are visible at a lower resolution.

To understand why, take a look at the cross-sectional image shown in Figure 50. As axons exit the gray matter, most of them undergo an important transformation — they become ensheathed by other cells that wrap around them repeatedly. Thus the brain not only wires itself up but also, amazingly, manages to wrap sheets of insulation around its "wires." The sheets are made of a substance called myelin, which is composed mostly of fat molecules. It's those molecules that make

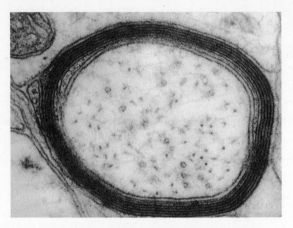

FIGURE 50. *Cross-section of myelinated axon*

white matter look white. (The epithet "fathead" may sound deroga-
tory, but it's actually accurate for everyone.) Myelination speeds up
the propagation of spikes, which is important for transmitting signals
quickly in large brains. Diseases of myelination, such as multiple scle-
rosis, have catastrophic effects on brain function.

The myelinated axons of the white matter are much thicker (typically 1 micrometer) than the mostly unmyelinated axons of the gray
matter. Furthermore, if we only care about finding regional connections, there's no need to see synapses. If an axon enters and branches
in a region of the gray matter, we can be almost certain that it makes
synapses there, so tracing the "wires" of the white matter is enough for
finding the regional connectome. If we restrict ourselves to myelinated
axons, we could accomplish the job with serial light microscopy, which
is similar to serial electron microscopy but employs thicker slices and
produces images with lower resolution.

Of course, mapping white-matter axons is still a daunting technical
challenge for a brain of human size. Studying white matter in smaller
brains, such as those of rodents and nonhuman primates, is a good
starting place. We can check the results by comparing them with those
from older techniques for studying white-matter pathways in animals.

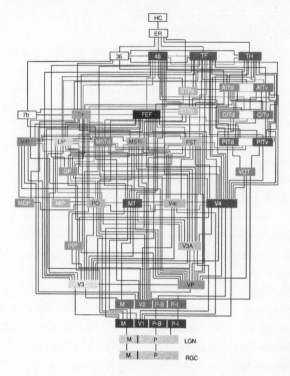

FIGURE 51. *Connections between visual areas of the rhesus monkey cortex (see Figure 39)*

These techniques were used to find connections between the visual areas of the monkey cortex, as shown in Figure 51. (The areas, but not their connections, were shown earlier.) Since the older techniques are not applicable to human brains, our own white matter has gone almost completely unexplored.

The Human Connectome Project is already trying to find a map like the one in Figure 51 for the human brain using diffusion MRI (dMRI) rather than microscopy. Diffusion MRI is different from MRI, which is used to find the sizes of brain regions, or fMRI, which is used to measure their activations. Unfortunately, dMRI is subject to the same basic limitation as other forms of MRI: poor spatial resolution. MRI typically yields millimeter-scale resolution, which is not enough for seeing single neurons or axons. Given its poor resolution, how can dMRI hope to trace the wires in the white matter?

It turns out that white matter has an interesting feature that makes its structure simpler than that of gray matter. Have you ever forgotten to stir the spaghetti after dropping it into boiling water? You discover your mistake a few minutes later, when you see that some of the strands have stuck to each other to form bundles. This culinary embarrassment resembles white matter; gray matter is more like a bowl of fully entangled spaghetti.

When axons bundle like unstirred spaghetti, they form a "fiber tract" or a "white-matter pathway." The bundles are similar to nerves, except that they run within the brain. Why do axons bundle? Well, why do so many people follow the same dirt paths through lawns? First, they are shortcuts, more efficient than the paved walkways installed by landscape designers. Second, there is a "follow the leader" effect — once a few trailblazers have worn down the grass a bit, everyone else follows them, trampling it down completely. Similarly, axons take efficient paths through the white matter, assuming that it evolved to achieve wiring economy. Since an efficient solution is often unique, we'd expect axons sharing the same origin and destination to take the same path. Also, it's known that the first axons to grow during brain development often blaze the trail, providing chemical cues for other axons to follow.

Fiber tracts may be thick, even though a single axon is microscopically thin. The largest is the famous corpus callosum, the huge collection of axons that travel between the left and right hemispheres. Neuroanatomists in the nineteenth century discovered other large tracts through naked-eye dissection of the brain. Diffusion MRI is an exciting advance, because it's a way of tracing white-matter pathways in the living brain. It computes an arrow at every location that indicates the orientation of the axons there. By connecting these arrows, it's possible to trace the paths of axonal bundles. In one notable success, dMRI has uncovered white-matter pathways connecting Broca's and Wernicke's regions, other than the classical one in the arcuate fasciculus. As I mentioned earlier, such discoveries are sparking revisions of the Broca–Wernicke model of language.

Such stories are encouraging, but dMRI also has limitations. Because of its poor spatial resolution, dMRI has difficulty following thin fiber tracts. And even thick tracts can be problematic if they intersect and their individual axons become intermingled. Think of this crossing as a chaotic traffic intersection packed with pedestrians, bicyclists, animals, and cars — you have to watch carefully to see whether any particular traveler goes straight or turns. Similarly, once axons enter the region where two bundles intersect, it's difficult to see, using dMRI, where they end up. The only foolproof way of mapping the white matter is to use a method that can trace individual axons, like the one I've proposed here.

Mapping regional connectomes is already problematic with dMRI; the method is even more ill-suited for neuronal and neuron type connectomes. Of course, dMRI has the important advantage that it can be performed on a living brain. At the very least it will detect gross connectopathies, like a missing corpus callosum. Since dMRI can be used quickly and conveniently to study many living brains, it will find correlations between mental disorders and brain connectivity. But these correlations might remain weak, just like the earlier phrenological ones.

MRI experts are continuing to improve resolution, but the rate of improvement is not that fast, and there is a long way to go. Roughly speaking, the current resolution of dMRI is a thousand times worse than light microscopy, which in turn is a thousand times worse than electron microscopy. Inventors might create better noninvasive imaging methods than MRI. But let's not forget that seeing through the skull into the interior of a living brain is fundamentally more challenging than chopping up a dead brain and examining the pieces with a microscope. Microscopy already delivers the resolution we need to find connectomes; we just have to scale it up to handle larger volumes. In contrast, MRI requires breakthroughs far more fundamental. For the foreseeable future, then, microscopy and MRI will remain complementary methods.

To find connectopathies, we will use the methods I outlined above to map reduced connectomes of abnormal and normal brains, and compare them. Some differences may be detectable by dMRI, but subtle ones will require microscopy. We will also compare neuronal connectomes of small chunks of brain using electron microscopy. The use of microscopy poses difficulties, as it must be carried out on the brains of the deceased. People do bequeath their brains to science — there is a long tradition of such generosity — but even if we have postmortem brains, many of them present special problems.

One alternative is to search for connectopathies in the brains of animals. Such research will also be important for developing therapies, which are often tested first on animals and only later on humans. The legendary French microbiologist Louis Pasteur produced the first vaccine for rabies by growing the virus in rabbits and then weakening it. The vaccine was tested on dogs before its dramatic first human trial on a nine-year-old boy who had been bitten by a rabid dog.

Studying human mental disorders with animals is no easy task. The rabies virus leads to the same disease, whether it infects rabbits, dogs, or humans. But is there such a thing as an autistic or schizophrenic animal? It's not clear whether such animals occur naturally, but researchers are now attempting to create them using the methods of genetic engineering. Researchers insert the faulty genes associated with autism and schizophrenia into the genomes of animals, usually mice, with the expectation of giving them analogous disorders. Ideally, such animals would serve as "models" for human disorders, approximations to the real thing.

But this strategy, a variation on Pasteur's, sometimes fails even for infectious diseases. Human immunodeficiency virus (HIV), which causes AIDS in humans, fails to infect many primates, making it difficult to test HIV vaccines. In monkeys, AIDS is caused by simian immunodeficiency virus (SIV), which is related to HIV but not identical.

The lack of a good animal model for human AIDS has slowed down research on finding a cure. Likewise, inserting faulty human genes into animals might not give them autism and schizophrenia. Some analogous but different genetic defect might be necessary.

Because of these uncertainties, the problem of validating animal models for mental disorders has risen to the fore. It's not clear what criteria should be used. Some emphasize similarity of symptoms, but even for infectious diseases this criterion doesn't always work. Sometimes the same microbe can infect both animals and humans but produces very different symptoms. An animal might tolerate infection with little adverse effect at all. And if human genes for autism or schizophrenia turned out to produce very different symptoms in mice, it wouldn't necessarily mean that the mouse models were useless. (Some might argue that it's pointless to compare symptoms, as mental disorders involve behaviors that seem uniquely human.)

An alternative criterion is similarity of neuropathologies, already being applied to evaluate mouse models of neurodegenerative disorders such as Alzheimer's disease (AD). In humans, AD is accompanied by abnormal buildup of plaques and tangles in the brain. Normal mice do not develop AD, but researchers have genetically engineered several mouse models that do. Their brains generate large numbers of plaques and tangles. Researchers are still arguing about whether any of these models are good enough for studying AD. But at least they have a target: a clear and consistent neuropathology to emulate.

Along these lines, similarity of connectopathies might be a good criterion for animal models of disorders like autism and schizophrenia. Of course, for this to work we would have to identify connectopathies in animal models, as well as analogous ones in patients afflicted by autism and schizophrenia.

You may have noticed that the plan for comparing connectomes sounds very different from the plan for decoding them. The connectionist theory of memory proposes particular hypotheses—the cell assembly and the synaptic chain—that can be tested using connectomics. In contrast, the connectopathy idea is open-ended. Without specific hypotheses, wouldn't searching for connectopathies be a wild-goose chase?

One of the leaders of the Human Genome Project, Eric Lander, has summed up the decade since its completion in this way: "The greatest impact of genomics has been the ability to investigate biological phenomena in a comprehensive, unbiased, hypothesis-free manner." It doesn't sound like what we were taught about the scientific method in school, where we learned that science proceeds in three steps: (1) Formulate a hypothesis. (2) Make a prediction based on the hypothesis. (3) Perform an experiment to test the prediction.

Sometimes that procedure works. But for every success story, there are many more stories of failure caused by choosing the wrong hypothesis to investigate. It can take a lot of time and effort to test a hypothesis, which might turn out to be wrong or — even worse — simply irrelevant. In the latter case, it would lead to research that ends up being a complete waste of time. Unfortunately, there's no well-defined recipe for formulating hypotheses, beyond a stroke of insight or inspiration.

We do have an alternative to "hypothesis-driven," or deductive, research — the "data-driven," or inductive, approach. It too has three steps: (1) Collect a vast amount of data. (2) Analyze the data to detect patterns. (3) Use these patterns to formulate hypotheses.

Some scientists gravitate to one approach over the other, because it fits their personal style. But the two approaches are not really in opposition. The data-driven approach should be viewed as a way of generating hypotheses that are more likely to be worth exploring than ones based purely on intuition. It can be followed by hypothesis-driven research.

If we have the right technologies, we'll be in a position to apply this approach to mental disorders. Connectomics will provide more and more accurate and complete information about neural connectivity. With so much data available, we'll no longer have to search for our keys under the lamppost. Once we identify connectopathies, these will suggest good hypotheses about the causes of mental disorders that are worth exploring further.

To resort to another metaphor, searching for the causes of mental disorders is like looking for a needle in a haystack because the brain is so complex. How to succeed? One way is to start from a good hypothesis about the location of the needle. Then you need search only a small part of the haystack. This will work if you are lucky or smart enough to have a good hypothesis. Another way is to build a machine that rapidly sifts through all the material in the haystack. You are guaranteed to find the needle with this technology, even if you're not lucky or smart. This is analogous to the connectomic approach.

To understand why minds differ, we have to see better how brains differ. That's why comparing connectomes is so crucial. Uncovering just any kind of difference won't be sufficient, however, since many differences could end up being uninteresting. We'll have to narrow in on the important ones, those that are *strongly* correlated with mental properties. These are the differences that will finally give connectionism more explanatory power than phrenology. They will accurately predict mental disorders for *individuals,* as well as faithfully estimate the intellectual abilities of normal people. (For connectomes obtained using microscopy on dead brains, the test would actually involve "postdiction," guessing the mental disorders or abilities of the deceased from their brains.)

Identifying connectopathies will be an important step toward understanding certain mental disorders. But understanding goes only so far. Ideally we will capitalize on it by developing better treatments for these maladies, or even cures. In the next chapter I'll envision how this will be done.

THIRTEEN

Changing

N 1821 THE composer Carl Maria von Weber premiered his opera *Der Freischütz.* To marry Agathe, the hero, Max, must impress her father by prevailing in a shooting contest. Driven to desperation by fear of losing his love, he sells his soul to the devil for seven magic bullets, which are guaranteed to hit their mark. Max not only wins the hand of Agathe but manages to evade the devil, and the opera ends happily.

In 1940 Warner Bros. released *Dr. Ehrlich's Magic Bullet,* which dramatized the life of the German physician and scientist Paul Ehrlich. After sharing a 1908 Nobel Prize for his discoveries about the immune system, Ehrlich didn't rest on his laurels. His institute discovered the first antisyphilis drugs, relieving the suffering of millions of people. By creating the first man-made drugs for any disease, Ehrlich effectively invented the entire pharmaceutical industry. He was guided by his theory of the "magic bullet," the name of which may have been inspired by Weber's popular opera. Ehrlich first imagined — and then discovered — chemicals that killed bacteria but spared other cells, like a magic bullet that unerringly flew to its target.

The bullet metaphor illustrates two important principles that apply to all medical treatments, not only drugs. First, there should be a specific target, and second, the ideal intervention should selectively affect *only* that target — that is, avoid "side effects." These principles aren't upheld by our remedies for brain disorders, which remain distressingly primitive. The surgeon's knife seems hopelessly crude for altering the brain's intricate structure, yet sometimes there is no other way. You've heard that neurosurgeons treat severe cases of epilepsy by removing the part of the brain where the seizures originate. But overzealous surgery can lead to catastrophe, as you saw in the case of H.M. To minimize side effects, it's important to target as small a region as possible.

Epilepsy surgery simply removes neurons from a connectome. Other procedures are intended to break the wires of neurons without killing them. In the first half of the twentieth century, surgeons attempted to treat psychosis by destroying the white matter connecting the frontal lobe to other parts of the brain. The infamous "frontal lobotomy" was eventually discredited and replaced by antipsychotic drugs. Yet psychosurgery is still practiced today as a last-ditch measure when other therapies fail.

Before considering other types of interventions, I'd like to step back to imagine the ideal one. I've said that certain mental disorders might be caused by connectopathies. If that's the case, true cures would require establishing normal patterns of connectivity. You might regard this prospect as hopeless if you're a connectome determinist. But even if you're more optimistic, you can't deny that the complexity of the brain's structure is daunting. Merely seeing connectomes is difficult enough, and repairing them seems even harder. It's unclear how any of our technologies could be up to the challenge.

But the brain is naturally endowed with mechanisms for connectome change — reweighting, reconnection, rewiring, and regeneration — that are exquisitely controlled. Since genes and other molecules guide the four R's, they could serve as targets for drugs. I doubt you're surprised by the idea of the connectome as the target for medications, given that you've been reading this book. But you might wonder whether the idea is consistent with what you know from other sources.

According to well-known theories dating back to the 1960s, certain mental disorders are caused by surplus or deficiency of neurotransmitter, which explains why they are relieved by drugs that alter neurotransmitter levels. Depression, for example, has been attributed to a dearth of serotonin, which is thought to be corrected by antidepressant medications such as fluoxetine, more commonly known as Prozac. (The drugs are supposed to increase serotonin levels by preventing neurons from sucking the molecule back up after secreting it. Recall that a number of such housekeeping mechanisms exist for keeping neurotransmitters from lingering in the synaptic cleft.)

But there is a problem with this theory. Fluoxetine affects serotonin levels immediately, yet it lifts mood only after several weeks. What could account for this long delay? According to one speculation, the serotonin boost causes other changes in the brain over the longer term. Perhaps it's these changes that relieve depression, but what exactly could they be? Neuroscientists have looked for effects of fluoxetine on the four R's, and found that it increases the creation of new synapses, branches, and neurons in the hippocampus. Moreover, as I mentioned in the discussion of rewiring, fluoxetine restores ocular dominance plasticity in adults, possibly by stimulating cortical rewiring. This doesn't prove that the drug's antidepressant effects are caused by connectome change, but it has certainly opened the minds of neuroscientists to the idea.

In this chapter I will focus on the prospect of finding new drugs that specifically target connectomes for the treatment of mental disorders. Let me emphasize, though, that other types of treatment are also important. Drugs may only increase the *potential* for change. To actually bring about positive changes, drugs could be supplemented by training regimens that correct behaviors and thinking. This combination could direct the four R's to reshape connectomes for the better. In my opinion, the best way to change the brain is to help it change itself.

• • •

There's no doubt that drugs have greatly advanced the treatment of mental disorders. Antipsychotics treat the most dramatic symptoms of schizophrenia, the delusions and hallucinations. Antidepressants can enable the suicidal to lead normal lives. But current drugs have limitations. Can we find new ones that are even more effective?

Our most successful drugs are for infectious diseases. An antibiotic like penicillin cures infections, killing bacteria by punching holes in their outer membranes. A vaccine consists of molecules that make the immune system more vigilant against a bacteria or virus. In short, an antibiotic corrects infection, while a vaccine prevents it.

These two strategies also apply to brain disorders. Let's consider prevention first. During a stroke, most neurons remain alive but damaged, and only later do they degenerate and die. Neuroscientists are working to find "neuroprotective" drugs that would minimize damage to neurons right after a stroke and thereby prevent death later on. The same strategy extends to diseases that destroy neurons for no apparent reason. For example, no one knows for sure why dopamine-secreting neurons degenerate and die in Parkinson's disease. Researchers hypothesize that the neurons are under some sort of stress, and would like to develop drugs that reduce it.

Some cases of Parkinson's disease are caused by defects in a gene that encodes a protein called parkin. An obvious therapy would be to replace the faulty gene. Researchers are attempting to do that by packaging a correct version inside a virus and injecting it into the brain, where they hope the virus will infect the dopamine-secreting neurons and protect them from degeneration. This "gene therapy" for Parkinson's has been tried in rats and monkeys so far, but not yet in humans.

Death is just the last step in the degeneration of a neuron, which is generally a long drawn-out process. You might compare it to the slow decline of a person who starts out weak and is then hit by a cascading progression of ailments, each worse than the last. To find clues, researchers look carefully at the various stages of degeneration in neurons, much as physicians observe the progression of symptoms in diseased patients.

Such observations are helpful because they narrow the search for molecular causes, the potential targets for neuroprotective drugs. In addition, they pinpoint the very first steps of degeneration. The timing is critical; intervening at the outset is likely to be more effective at preventing cell death later on. Early intervention is also important for treating cognitive impairments, which often emerge long before significant neuron death. These symptoms may occur because connections are lost well before neurons actually die.

In general, it's important to see degeneration more clearly, and to see it at its earliest stage. The images acquired by the tools of connectomics will help us do that. Serial electron microscopy will reveal exactly how a neuron deteriorates. We will also obtain more precise information about which neuron types are affected and when. All this is bound to be helpful in the search for ways to prevent neurodegeneration.

Can we also find ways to prevent neurodevelopmental disorders? To do this, we must diagnose them as early as possible, before development has veered too far off course. Even while the fetus is still in the womb, genetic tests can be performed to predict whether disorders such as autism and schizophrenia are likely to emerge later on. But accurate predictions may require combining genetic testing with examination of the brain.

I argued earlier that microscopy of dead brains, with its high spatial resolution, will be necessary for determining whether a brain disorder is caused by a connectopathy. That method might yield good science, but by itself it will be useless for medical diagnosis. That being said, once a connectopathy has been fully characterized by microscopy of dead brains, it should become easier to use diffusion MRI to diagnose it in living brains. In general, it's easier to detect something if you know exactly what you are looking for.

Behavioral signs will also be informative for some disorders. Some schizophrenics exhibit mild behavioral symptoms when they are children, before the first onset of true psychosis. Perhaps careful detection of such early symptoms, combined with genetic testing and brain imaging, could accurately predict schizophrenia.

Early diagnosis of neurodevelopmental disorders will pave the way for prevention. Connectomics will help us identify exactly which processes of brain development are involved, making it easier for us to develop drugs or gene therapies that prevent connectopathies or other abnormalities from developing.

The goal of prevention seems ambitious enough; it's even more challenging to repair the brain when the damage has already been done. After injury or degeneration has caused neuron death, is there any recourse? A pessimistic answer comes from regeneration denial, one flavor of connectome determinism. Since it's generally true that no new neurons are added in adulthood, the brain has limited power to heal itself after injury. Is there any way to overcome this?

Other animal species, such as lizards, are able to regenerate large parts of their nervous systems after injury. And human children regenerate better than adults do. In the 1970s, when physicians realized that children's fingertips regenerate like lizards' tails, they stopped attempting to reattach severed fingertips through surgery; now, they simply let the fingertips grow back. Hidden powers of regeneration might lie dormant in adults, and the new field of regenerative medicine seeks to awaken them.

Injury naturally activates regenerative processes in the adult brain. A main site of neuron creation is known as the subventricular zone. Immature neurons, known as neuroblasts, normally migrate from there to the olfactory bulb, a brain structure dedicated to smell. Stroke increases the creation of neuroblasts and can divert them from the bulb to the injured brain region. Since this natural process might contribute to recovery after stroke, some researchers are trying to develop artificial means to promote it.

Another route to regeneration is to transplant new neurons directly into the damaged region. This might work better than trying to promote migration from a distant location like the subventricular zone. Parkinson's disease, as I've mentioned, involves the death of dopamine-secreting neurons. Researchers have attempted to replace them by transplanting healthy neurons from fetuses. Amazingly, some neurons were shown to survive in recipients' brains for over a decade, although it's unclear whether the transplants actually did much to alleviate the symptoms. The experiments, conducted with cells isolated from aborted fetuses, raised thorny ethical issues. A further complication of transplantation was that patients' immune systems could reject the new cells as foreign.

We can now avoid both of these problems, thanks to a recent advance that allows the culturing of new neurons customized to a particular patient. A skin cell can be "deprogrammed" to become a "stem cell," one that has effectively "forgotten" its former life as a skin cell. Owing to its newly ambiguous identity, this stem cell can now be "reprogrammed" to divide and produce neurons in vitro. (The Latin term *in vitro*, which means "in glass," refers to the artificial environment used for culturing molecules, cells, or tissues isolated from an organism. At first that environment was typically a glass container, but plastic ones are more common now.) Researchers have used this method to create dopamine-secreting neurons from the skin cells of Parkinson's patients. They are planning to transplant the neurons back into the patients' brains to treat them.

Whether created naturally or added by transplantation, most new neurons die. Without "taking root," new neurons presumably cannot survive. Regenerative treatments will thus require enhancing the integration of new neurons into the connectome, a process that depends on promoting the other three R's — rewiring, reconnection, and reweighting.

The adult brain may hold untapped potential for making these changes. Earlier I referred to the fact that most recovery happens during the three-month period just after stroke. According to one speculation, this is a critical period, analogous to the one during brain development, with production of similar molecules that promote plasticity. Once this window closes, plasticity plummets and the rate of recovery slows. Perhaps stroke therapies should aim at keeping the window open, extending the natural processes of recovery.

As we've seen, rewiring may be difficult in the adult brain. After injury, though, neurons appear to grow new axonal branches more easily. If researchers can identify the molecular reasons why, it may be possible to promote rewiring of the adult brain by artificial means, which would help integrate new neurons into the brain as well as allow existing neurons to change their functions. Similarly, since creation of new synapses happens at a greater rate in the injured brain, there may be natural molecular processes that could be manipulated to promote reconnection.

Could we also correct neurodevelopmental disorders, fixing the brain after it has wired up improperly? If you're a connectome determinist, you'd probably regard correction as futile and instead focus all your efforts on prevention. But it's not clear whether completely accurate and early diagnosis of neurodevelopmental disorders will be possible, so we have no choice but to think about correction too. This will require the most extensive connectome changes of all, and therefore the most advanced control of the four R's.

I've stressed the treatment of malfunctioning brains, since these are the connectomes most in need of change, but people also want drugs for enhancing normal brain function. Many university students drink coffee while studying. While caffeine may help them stay awake, it has little effect on learning and memory. Nicotine improves the mental abilities of smokers, but that's only relative to their substandard performance when deprived of cigarettes. Can we find more effective drugs than these? For example, we'd really like a drug that promotes the connectome changes necessary for learning or remembering new information or skills. Also useful would be drugs to help us forget. Perhaps these could promote the elimination of cell assemblies or synaptic chains formed after traumatic events, or those implicated in bad habits or addictions.

We have a long wish list for drugs, both for preventing brain disorders and for correcting them. Unfortunately, the pace of discovery is slow. New drugs appear on the market every year, often with great fanfare, but many are not really new; they're just variants of old drugs, and unlikely to be significantly more effective. Most antipsychotics and antidepressants are variants of drugs discovered by accident over half a century ago. Few drugs are truly new; few draw on recent advances in neuroscience.

The challenges of drug development are not unique to mental disorders, of course. Creating new pharmaceuticals is a hugely risky business. It can take many years to develop candidate drugs. Only those deemed most likely to succeed are tested in human patients, yet nine out of ten fail in this last stage, turning out to be toxic or ineffective. This is a huge waste of money, given that clinical trials incur a significant fraction of the investment required to bring a new drug to the marketplace. (Total cost estimates range from one hundred million to a billion dollars.) Everyone desperately wants better drugs — those who suffer from diseases, those who treat them, and those who invest huge sums of money trying to develop therapies. How can drug discovery be accelerated?

Historically, most drugs have been discovered by chance. The first antipsychotic was chlorpromazine, known in the United States by the trade name Thorazine. This belongs to the phenothiazine class of molecules, the earliest of which were originally synthesized in the nineteenth century by chemists attempting to create dyes for the textile industry. In 1891 Paul Ehrlich discovered that one of them could be used to treat malaria. During World War II, the French pharmaceutical company Rhône-Poulenc (a forerunner of today's Sanofi-Aventis) tested many phenothiazines looking for more malaria drugs; when they failed to find any effective ones, they started looking for antihistamines. (You may have taken medications of this type for allergies.) Then a physician discovered that phenothiazines could enhance the actions of surgical anesthetics. Rhône-Poulenc researchers switched to testing for this new application, and discovered that chlorpromazine was effective. After giving the drug to psychiatric patients as a sedative, doctors realized that it specifically reduced symptoms of psychosis. By the end of the 1950s, chlorpromazine had swept through the psychiatric hospitals of the world.

The first antidepressant medications, iproniazid and imipramine, were discovered around the same time, in stories with similar twists and turns. Iproniazid was originally developed for tuberculosis, but had the unexpected side effect of making patients unreasonably happy. Psychiatrists eventually realized that it could be used to treat victims of depression. Meanwhile, the Swiss company J. R. Geigy (an ancestor of Novartis), having heard about Rhône-Poulenc's success with chlorpromazine, decided to play catch-up by looking for an antipsychotic of their own. They tried testing imipramine, which chemists had synthesized by modifying a phenothiazine. It was a failure for treating psychosis, but fortunately it turned out to relieve depression.

So researchers did not intend to develop the first antipsychotics and antidepressants. They were just lucky and alert enough to stumble upon them during this golden age of the 1950s. More recently, there has been growing excitement about "rational" methods of drug discovery built upon our modern understanding of biology and neuroscience. How do these methods work?

Recall that cells are composed of a huge variety of biological molecules, which are involved in many kinds of life's processes. (Earlier I talked about the important class of biomolecules known as proteins, which are synthesized based on blueprints encoded in genes.) A drug is an artificial molecule that interacts with the natural ones in cells. Ideally, according to the magic-bullet principle, the drug should interact with a specific type of biomolecule but not with other types.

Rational drug discovery, therefore, starts from biomolecules involved in the processes that malfunction during disease. Researchers have begun to identify many such biomolecules, which can serve as targets for therapies. The tempo of target identification has quickened with the advent of genomics, engendering increasing optimism about finding new drugs by rational means.

Once a drug target is identified, the first task is to find artificial molecules that bind to it, like a key fitting into a lock. Researchers create a variety of candidates, based on educated guesses, and proceed to test them empirically. If they manage to hit the target with a candidate, they refine its structure, progressively improving its binding with the target. This first stage of drug development is conducted by chemists.

Let's jump ahead for the moment to the last stage, human testing. Physicians manage this stage, administering candidate drugs to patients to see whether symptoms improve. It's neither economical nor ethical to test a drug on people unless there is already good reason to believe that the drug is likely to be safe and effective. Even so, nine out of ten candidates fail at this point, as I mentioned earlier, and the attrition rate is even higher for disorders of the central nervous system. These depressing statistics suggest that something is going wrong between the first and last stages of drug development. Before commencing human testing, how can researchers be more certain that a candidate drug will not only bind to its target biomolecule in vitro but also be effective at treating the disease? Finding more evidence, or evidence that's more reliable, would make it faster and cheaper to develop new drugs.

One method is to test in animals first, but it's even more difficult to create animal models for mental disorders than for other kinds of diseases. As I've mentioned, researchers are using the genetics of autism and schizophrenia to develop mouse models. But mice may not be enough like humans to have these disorders, so some researchers are planning to develop models based on nonhuman primates.

Drugs can also be tested on in vitro disease models. One exciting approach is based on the "stem cells" that can be created from a patient's skin cells and reprogrammed to divide into neurons. Earlier I described the plan to transplant these neurons back into the patient's brain to treat neurodegenerative disorders. Another option is to keep such neurons alive in vitro and use them for drug testing. Cultured neurons generate spikes and transmit messages through synapses, much as in the brain, and hence can be used to assay the effects of drugs on these functions. These neurons wire up very differently from those inside the brain, however, so in vitro models might not be useful for mental disorders that are caused by connectopathies.

Finally, it's possible to "humanize" animal models by growing human neurons from stem cells and then transplanting them into animal brains. This might yield better animal models than the approach based on inserting defective human genes. Researchers are already adopting similar strategies to create humanized mouse models for diseases other than mental disorders.

Along with creating better in vitro and animal models, we must also figure out how to evaluate success when testing candidate drugs on them. The obvious approach for animal models is to administer the drug and then quantify the resulting changes in behavior. To do this, we need to observe some animal behavior that is analogous to a symptom of the human mental disorder. But it's no easy task to define such behaviors. (What exactly is a psychotic mouse?) That's why it's not so obvious how to evaluate drugs with tests of animal behavior.

Could there be some other way? Drugs for neurodegenerative diseases such as Parkinson's can be tested for their effectiveness in preventing the death of neurons in animal models of these diseases. Likewise, it might be better to evaluate drugs for autism and schizophrenia by looking at their effects on neuropathologies rather than behavioral symptoms. But this approach has been blocked by the failure to identify clear and consistent neuropathologies. If autism and schizophrenia turn out to be caused by connectopathies, it will be important to identify analogous miswiring in animal models. Then drugs could be tested for their effectiveness in preventing or correcting such miswiring. To make this approach practical, we'd have to speed up the technologies of connectomics in order to compare many animal brains quickly.

Earlier I claimed that studying mental disorders without connectomics is like researching infectious diseases without a microscope. My claim extends to research on treatments. If you can't even see a connectopathy, it's bound to be difficult to find therapies that prevent or correct it. Furthermore, research on the molecules involved in the four R's of connectome change is likely to be a prime avenue for identifying drug targets. I expect that connectomics will play a central role in developing psychiatric therapies, much as genomics has already taken center stage in pharmaceutical research more generally.

• • •

Curing mental disorders sounds like a worthy goal. So also does re-wiring the brain of a soldier traumatized by war, or a child who has suffered severe abuse. Yet the means I've discussed, manipulating the genes and neurons of animals and humans, might provoke a twinge of fear. Anxiety over biotechnology dates back a long time. In his 1932 novel *Brave New World*, the English writer Aldous Huxley imagined a future dystopia based on transforming the body and brain. Humans are born in factories controlled by the state, separated into five bio-logically engineered castes, and provided with a mind-altering drug called "soma" in place of religion.

While we should be vigilant against potential abuses of biotechnol-ogy, I don't think we should be fearful. Because of their complexity, living systems have proven quite difficult to reengineer. It's not impos-sible to do so, but it generally takes longer than alarmists anticipate. Progress happens slowly, which gives human societies ample time to figure out how to handle it.

Optimism about biotechnology is as old as pessimism. A contem-porary of Huxley's, the Irish-born biologist J. D. Bernal, presented an upbeat view in his 1929 essay "The World, the Flesh, and the Devil." He saw humanity's story as a quest for three types of control. Power over "the world" was already growing — this was the goal of the physi-cal sciences and engineering. Control over "the flesh" seemed farther off, but Bernal predicted that future biologists would learn to manipu-late genes and cells. His most prophetic remarks were reserved for the third challenge:

> Why do the first lines of attack against the inorganic forces of the world and the organic structure of our bodies seem so doubtful, fan-ciful and Utopian? Because we can abandon the world and subdue the flesh only if we first expel the devil, and the devil, for all that he has lost individuality, is still as powerful as ever. The devil is the most dif-ficult of all to deal with: he is inside ourselves, we cannot see him. Our capacities, our desires, our inner confusions are almost impossible to understand or cope with in the present, still less can we predict what will be the future of them.

Bernal feared that our mental flaws ("the devil") would be the ultimate barrier to our progress. The third and final challenge for humanity was to reshape the psyche.

Would Bernal be happy to see how far we've come? We have survived the threat of annihilation by nuclear weapons (until now, anyway). Perhaps we have learned enough that we will never again wage wars as terrible as those of the twentieth century. But Bernal would note that we struggle more than ever to deal with the consequences of our desires. Our control over "the world" has made inroads on the problem of scarcity, but abundance turns out to be dangerous too. Our lack of self-control drives us to pollute the environment and sicken our bodies with overconsumption.

Perhaps we can resist "the devil" by restructuring our economic incentives, reforming our political systems, and perfecting our ethical ideals. These are the time-honored ways of improving our brains. But in time, science will also invent others. Bernal hoped that humanity would triumph over the world, the flesh, and the devil, which he called "the three enemies of the rational soul." We can express his dream in another way — as the quest to control atoms, genomes, and connectomes.

"Science is my territory," the physicist Freeman Dyson wrote, "but science fiction is the landscape of my dreams." In the final part of this book, I'll look at two fantasies from our collective dream landscape: cryonics, the practice of freezing a corpse in the hope that it will be resurrected later on by some advanced civilization, and uploading, the idea of living happily ever after as a computer simulation.

Bernal opened his essay with an oracular pronouncement: "There are two futures, the future of desire and the future of fate, and man's reason has never learnt to separate them." Since many people wish to live forever, we should be skeptical of cryonics and uploading. Mere wishful thinking is the "future of desire," a mirage that distracts us from the "future of fate." To examine these dreams critically, we must reason rather than wish, and our thinking will inevitably turn to connectomes.

PART V

BEYOND HUMANITY

FOURTEEN

To Freeze or to Pickle?

T WICE IN MY life I have visited that strange town in the desert called Vegas. Each morning, I luxuriated in the soft sheets of my hotel bed. Each night, glittering spectacles of entertainment held me in thrall. I savored shots of whiskey and blew cigar smoke toward the lofty ceiling of the casino. But the blackjack table and the roulette wheel left me bored and listless.

Games of chance cannot hold my attention — save one, the only gamble that really matters. It is called Pascal's Wager. In 1654 the French genius Blaise Pascal founded the branch of mathematics known as probability theory. That same year, he also found God. After a searing religious vision, the focus of his life shifted from science and mathematics to philosophy and theology. His most important work during this period was a defense of Christianity, which was still unfinished when he died prematurely at age thirty-nine. His notes were published posthumously under the name *Pensées* (Thoughts). We encountered the *Pensées* at the beginning of this book. Now we return to them as we near its conclusion.

The *Pensées*, as you might have guessed from the passage I quoted earlier, are full of dread. To Pascal, dread was not a nihilistic end in itself; it was a prelude to religious faith. Pascal was well aware that the greatest affliction of the believer is doubt. How can we be sure that God exists? Many philosophers and theologians had argued that the existence of God can be proven by logic and reason. Pascal, though familiar with their purported proofs, was not convinced.

So he proposed a radically different approach. He gave up trying to banish skepticism, and granted that a rational person could never be certain of God's existence. One could only estimate the *probability* that God exists. Even so, Pascal argued, it made sense to believe in God. His creative stroke was to formulate faith as a gamble. You are faced with two choices: Believe or not believe. There are two possible realities: Either God exists or he does not. The table in Figure 52 shows the four possible outcomes.

		GOD	
		exists	doesn't exist
YOU	believe	Whew! gain all	Oh Well... lose little
	don't believe	Damn! lose all	Fun gain little

FIGURE 52. *Pascal's Wager*

On the one hand, if you don't believe in God, you'll get to partake of the sinful pleasures that the nuns in your Catholic school taught you to resist. But you'll also have to risk burning in hell for all eternity. On the other hand, suppose you choose to believe in God. There are costs to belief, such as having to sit on those uncomfortable church pews every Sunday morning when you'd rather be sleeping or playing tennis. But it might be worthwhile if God exists, for then you would receive the fantastic prize of eternal life in heaven.

The table indicates the reward or punishment of each possible outcome. If you were mathematically minded, you would fill in the table with numbers that quantify how much you dislike church, or how hellish you imagine hell to be. You'd also have to estimate the probability that God exists, thus quantifying your skepticism or belief. Then you'd calculate the expected payoffs from believing and not believing, and choose accordingly.

But Pascal saved us from having to calculate so diligently, by pointing out that the result is obvious without our actually doing the numbers. The value of heaven is infinite, since eternal life is infinitely long. Infinity times any number is still infinity. Therefore, the expected payoff from believing in God is infinite, as long as the probability of God's existence is any number greater than zero. The precise values of the other numbers don't matter at all. In short, going to church is like purchasing a lottery ticket. It's worth paying any price for the ticket if the jackpot is infinite.

Centuries have passed since Pascal. Times have changed, and the new millennium has given rise to a new wager. To see the modern gamblers, we must journey to Scottsdale, Arizona, in search of a strange warehouse. Entering the building, we see rows of metallic containers, each a bit taller than a human. The containers are called dewars, and like giant thermos bottles they insulate their contents. Instead of holding a refreshing drink for a summertime hike, the dewars store liquid nitrogen, and instead of ice cubes they contain either four human corpses or six human heads.

This is the headquarters of the Alcor Life Extension Foundation. The foundation has about one thousand living members and one hundred dead ones. You can join the club by guaranteeing a $200,000 payment, to be handed over when you are pronounced legally dead. In return, the foundation promises to preserve your body indefinitely at 196 degrees below zero. (All temperatures are in the Celsius scale.) You can opt to preserve only your head, in which case the price drops to $80,000. The foundation has its own language. The people inside the dewars are not dead; they are "deanimated." The frozen heads are "neuropreservations," and the practice is called "cryonics."

Alcor members are optimists, as is clear from a twenty-eight-minute promotional video called *The Limitless Future*. In the long run, advances in science and technology will enable humans to accomplish what seems impossible today. Humankind's ability to control matter will become so sophisticated that it will eventually be possible to "re-animate" dead bodies. Not only will the frozen corpses in the Alcor warehouse be brought back to life, but their diseases and old age will be reversed. The reanimated will be restored to their youthful vigor.

The physicist Robert Ettinger was the first to bring the idea of cry-onics before the attention of the public. Through television appear-ances and his best-selling 1967 book, *The Prospect of Immortality,* he became a minor celebrity. All the same, it took a number of years and several false starts before cryonics became a reality. In the early years there were some embarrassing episodes in which frozen bodies ac-cidentally thawed and had to be buried just like the corpses of other dead people. Finally, in 1993, the Alcor Life Extension Foundation cre-ated the facility in Scottsdale, which seems secure enough to keep bodies frozen for many years.

Ettinger was successful in popularizing his ideas, but he was ridi-culed as well. Indeed, it's tempting to dismiss Alcor members as suck-ers who have been fleeced of large sums of money. But this reaction would be too hasty. Can anyone really prove that reanimation will al-ways be impossible? It seems more reasonable to say that the probabil-ity of reanimation is small but nonzero. This opens the door to Pas-calian arguments. The expected value of Alcor membership is equal to the probability of reanimation times the value of eternal life. Since eternal life is infinitely valuable, the expected value of Alcor member-ship is also infinite, and therefore worth every penny of the $200,000 fee. Like Christianity, cryonics is a wager for the jackpot of eternal life. Pascal's Wager asks you to put your faith in God; Ettinger's Wager asks you to put it in technology.

The twentieth-century French author Albert Camus opened his es-say *The Myth of Sisyphus* with a provocative claim: "There is but one truly serious philosophical problem, and that is suicide." I counter that there is only one truly serious problem in science and technology, and that is immortality. Through his dramatic opening, Camus introduced the question of whether life is worth living, whether life has meaning.

It's worth noting that suicide is a purely philosophical problem because there are no practical barriers. If you want to kill yourself, you are in luck — it's easy to find a gun, a rope, a tall building, or poison. But immortality is a technological problem. Even if you want to live forever, there is no option currently available.

The quest for eternal youth is as old as humankind. My schoolteachers told me that it was while searching for the Fountain of Youth that the Spanish explorer Ponce de León discovered Florida. That charming story is now considered apocryphal, alas, but historians still seem to believe records of two expeditions commissioned by the Chinese emperor Qin Shi Huang to find the fabled elixir of life in the third century B.C. With a fleet of ships and a crew of three thousand boys and girls, the court sorcerer Xu Fu sailed the eastern seas for years without success, and never returned from his second expedition.

Today, the quest for immortality is alive and well. Salesmen peddle vitamins, anti-oxidants, and anti-aging creams. These and other modern elixirs of life have more to do with wishful thinking than with reality. But some people think science is at last on the verge of breakthroughs in life extension. In his book *Ending Aging,* Aubrey de Grey has set forth his ideas on "Strategies for Engineered Negligible Senescence" (SENS). He lists seven types of molecular and cellular damage that occur during aging, and predicts that science will eventually be able to prevent or repair each one. De Grey co-founded the Methuselah Foundation, which offers monetary prizes to researchers who can make mice survive for record-setting life spans.

On the one hand, there are genuine scientific investigations of aging and longevity taking place today, and I would be foolish to criticize this type of research. Although the field of life extension has its share of charlatans, that shouldn't deter real scientific investigation. Aging and death are fascinating problems, even if there are no immediate prospects of solution. And who knows? Given enough time, humankind might attain immortality.

On the other hand, I am skeptical of extreme optimism about such matters. In his book *Live Long Enough to Live Forever,* the inventor Ray Kurzweil predicts that immortality will be attained in the next few decades. If you can manage to live long enough to survive to that point, you will live forever. Personally, I feel quite confident that you, dear readers, will die, and so will I.

If you are a long-term optimist but a short-term pessimist, what should you do? Why not prepare for your demise by joining Alcor? Dunk your body into a liquid nitrogen time capsule, so that it might last the centuries or eons required for humanity to master not only the art of immortality but also that of resurrection. Cryonics is a temporary measure, practiced by forward-looking members of a civilization advanced enough to make liquid nitrogen, but not advanced enough to live forever.

By now everyone seems to have heard of cryonics. (Some people say "cryogenics," but this refers to the generic study of low temperatures, not to the bid for immortality.) The turning point of public awareness was probably 2002, when the baseball star Ted Williams died. His son and daughter by his third marriage sent the body to Alcor for preservation. His daughter by his first marriage filed a lawsuit, citing the request Williams made in his will for cremation. During the bizarre court battle that followed, Alcor sat on the sidelines waiting for the verdict while Williams's severed head and body, chilled but not frozen, sat in their warehouse. Ultimately, Alcor received the rest of their fee and laid the athlete's remains to rest in liquid nitrogen.

According to my reading of public opinion, people are now becoming more willing to at least entertain the claims of cryonics. Alcor members have gone further, believing fervently enough to invest money in freezing. Religion has long been successful at convincing people to believe in the incredible. In 1917, a crowd of seventy thousand gathered near the Portuguese village of Fatima to witness the sun change colors and dance wildly in the sky, while three shepherd children proclaimed their visions of the Blessed Virgin Mary and the rest of the Holy Family. Every year, millions of pilgrims now voyage to the site of the "Miracle of the Sun," which was officially acknowledged by the Roman Catholic Church in 1930.

The pollsters tell us that 80 percent of Americans believe in miracles. I have heard some Christians scoff at such stories, regarding belief in miracles as primitive and vulgar. But let's not forget that Christianity makes much ado about the most famous miracle of all, the resurrection of Jesus Christ. According to the Roman Catholic doctrine of transubstantiation, miracles continue to happen every Sunday in every church, when the wafer and the wine are transformed into the body and blood of Christ. If you are religious, it is rational and consistent to insist on the miraculous. What other evidence can you have for the existence of forces that are supernatural?

Today we have fallen in love with another source of miracles. In the days preceding June 29, 2007, thousands of fanatics throughout the United States assembled in front of shrines to the technology of Apple Incorporated. Within the first day and a half of the iPhone launch, 270,000 customers had converted. Millions more followed suit by the end of the year. In the frenzy leading up to the most anticipated new release of the decade, some bloggers had dubbed it the "Jesus phone."

Judging from the excitement that it inspired, the iPhone was clearly out of the ordinary. One might even call it a modern miracle. If you think that's hyperbolic, imagine how the iPhone would be viewed by someone who lived in the nineteenth century. According to Clarke's Third Law of Prediction, "Any sufficiently advanced technology is indistinguishable from magic." Through a steady stream of miracles, technology has persuaded us of its amazing powers. A new cult of technological optimism has deeply embedded itself in the zeitgeist.

John the Baptist told us that the Messiah would come and that the Kingdom of God was nigh. Technology's prophet is Ray Kurzweil, and its gospel is his 2005 book, *The Singularity Is Near.* I've already mentioned Moore's Law, which describes the exponential growth in computational power that has astounded us over the past forty years. By extrapolating this glorious past to the future, and to other technologies beyond computers, Kurzweil presents a vision of a future that knows no limits.

His boundless optimism reminds me of Leibniz, whose views on perception I talked about earlier. Leibniz taught that we live in the best of all possible worlds, a doctrine that he deduced from a simple argument: Since God is perfect and all-powerful, surely he would never create anything less than the best world. Leibnizian optimism is mainly remembered through its lampooning by the French philosopher Voltaire. In the satirical novel *Candide,* the learned Dr. Pangloss tries to convince the other characters of the world's perfection, seemingly unaware of the evil and mayhem that surrounds them wherever they go.

Of course we don't live in the best of all possible worlds, but just wait — technology will get us there. Such is the Panglossian promise of Kurzweil. The whiff of possibility has drawn people to cryonics. In my opinion, their suspension of disbelief is a sign that they accept mechanism. That's the philosophical doctrine that the body — and therefore the brain — is nothing more than a machine. Granted, our bodies are much more sophisticated than the machines we make, but in the end, mechanism says, there is no fundamental difference.

We have resisted the doctrine for a long time. Even in the nineteenth century, some biologists stuck to the idea of a "vital force" present in living organisms and absent from the laws of physics and chemistry. In the twentieth century, advances in the field of molecular biology pushed "vitalism" to the wayside. Many still cling to some form of dualism, the idea that mental phenomena depend on something nonphysical, such as the soul. But plenty of people have been convinced by the discoveries of neuroscience that there is no "ghost in the machine."

If the body is a machine, why can't it be repaired? That possibility doesn't seem to violate the laws of logic or physics, assuming that you accept the doctrine of mechanism. In *The Sword and the Stone,* his telling of the legend of King Arthur, T. H. White satirized totalitarian societies by describing a colony of ants living in a nest with every entrance decorated by the slogan "Everything not forbidden is compulsory." Kurzweil has updated Leibniz, telling us that "Everything possible is inevitable."

But as every inveterate dreamer hates to be reminded, there are lots of possibilities that we never end up pursuing. Any decision involves weighing costs and benefits. Reanimation may be possible, but at what cost? Yes, a human life is priceless — but what if no bank contains enough money to pay? For example, suppose that reanimation is possible in principle, but in practice would require more energy than exists in the known universe. At some point, the constraint of finite or expensive resources starts to matter.

The difficulty of reanimation also matters for Alcor members, because it determines their time horizon. A wonderful selling point of cryonics is that while submersed in liquid nitrogen you can wait for all eternity and never get bored. But can you count on your resting place to remain intact? What is the chance that Alcor will still exist by the time reanimation becomes feasible, if it takes a million years of technological progress to reach that point?

Some believers in cryonics may choose to turn a blind eye to practical considerations. But those who are skeptical by nature will have to consider Ettinger's Wager. Pascal argued that there's no need for calculation, because the wager is for an infinite jackpot. Yet in reality, nothing about our universe is truly infinite. The rational decision-maker must, in the end, actually perform the probabilistic calculations. Although no one really knows the numbers involved, at the very least an estimate can be made. Doing that in an informed way requires some study of the scientific and medical issues.

It's true that any machine can be kept running indefinitely by replacing its broken parts. In 2007 the world's oldest running car was auctioned off. "La Marquise," which has a steam engine rather than an internal combustion engine, was built in 1884 by De Dion, Bouton et Trépardoux, the largest automobile manufacturer in the world for a time. But the price the car fetched — $3.2 million — tells you just how rare it is for a very old car to be in working order. Automobiles are generally designed to last for about a dozen years of use. Beyond twenty-five years of age, a car is considered an antique. Maintaining it longer than that is not cost-effective if the only goal is transportation; replacement parts are expensive to make in small quantities and to install piece by piece. Keeping a car running forever is worth doing only for aesthetic or sentimental reasons.

Of course, there are better reasons to keep humans running. Sometimes bodies can be repaired by replacing parts at great cost. Organ transplantation is made possible by drugs that suppress the recipient's immune system, preventing it from attacking the donor organ. It would be better to avoid the immune reaction altogether by using an organ made of cells that are genetically identical to those of the recipient. Right now this is possible only when transplanting an organ from one identical twin to another. But tissue engineers have the dream of culturing organs in vitro, by growing cells on artificial scaffolding. If they are successful, it will become possible to take cells from someone, grow an organ from them, and transplant the cultured organ back into the person. No donor would be necessary.

Optimistic though we may be about future advances in organ replacement, there's a fundamental limitation: The brain is an organ that cannot be replaced. That's not a statement about the technical difficulty of a brain transplant. What I'm talking about is the issue of personal identity, illustrated nicely by the true story of Sonny and Terry.

In 1995 Sonny Graham received a heart donated by Terry Cottle, who had committed suicide. In a surprising turn of events, Terry's widow, Cheryl, married Sonny nine years later. Four years into their marriage, Sonny committed suicide in the same way that Terry had, shooting himself in the head. The tabloids went crazy with headlines like "Suicide Claims Two Men Who Shared One Heart."

Reporters and bloggers erupted with wild speculations and questions. Did the transplanted heart contain memories that made Sonny fall in love with Cheryl? Did it drive Sonny to suicide, just as it had done to Terry? The story became less mysterious when the police found that Cheryl had been married five times, reportedly driving all of her husbands to despair. After receiving Terry's heart, Sonny was still Sonny. His personal identity remained intact. It's doubtful that it was the transplanted heart that made Sonny fall in love with Cheryl. More likely, he was attracted to Cheryl because she was attractive. (After all, she did manage to secure five husbands.)

In contrast, let's consider a hypothetical *brain* transplant. The procedure is impossible today, but it makes for an interesting thought experiment. Suppose Terry's brain had been transplanted into Sonny's body. It would not make sense to say that Sonny had received Terry's brain, since the postsurgical Sonny would not be the Sonny his friends knew. If they asked, "Sonny, remember the time we . . . ?" they'd get a blank stare in return. We might say instead that Terry had received Sonny's body. In other words, we could call it a body transplant rather than a brain transplant. Then Cheryl's second encounter with a suicidal husband might have a different explanation.

The bizarre story of Sonny and Terry introduces an important point for cryonics: Preservation of the brain is the pivotal issue. Most Alcor members have chosen the cut-rate option of freezing only their heads, believing — presumably — that any future civilization advanced enough to resurrect them will be advanced enough to replace their bodies. But will this future civilization also be able to revive their frozen brains?

This question faces anyone deciding whether to engage Alcor's services, but I think it's profoundly interesting even for those who don't care a whit about Alcor. Reanimation is the ultimate challenge for the doctrine of mechanism. Philosophers can argue until they're blue in the face, and scientists can uncover all the evidence they want, but they can never completely convince us that the body and the brain are machines. The final proof will come only when engineers manage to construct machines that are just as complex and miraculous. Or when they can bring dead bodies and brains back to life by repairing them like cars.

In a more practical vein, we can view the Alcor question as an extreme version of one asked in hospitals. Friends and family of a patient lying in coma would like to know: Will she ever wake up? Like the brains of the comatose, Alcor's brains have been damaged. Both types of brains blur the line between life and death. What are the fundamental limits of restoring life to damaged brains? Once again, we cannot properly address this question without considering connectomes.

Alcor's procedures are based on a field of science known as cryobiology. You probably know that fertility doctors freeze sperm, eggs, and embryos for later use. Blood banks freeze rare blood types for transfusion years later. The classic method is to lower the temperature slowly, say one degree per minute, after immersing cells in glycerol or other cryoprotective agents that increase their survival rate. The method is far from perfect. Sperm and embryos survive the best; eggs do less well. Cryobiologists would like to freeze entire organs, since it is wasteful to discard them just because immediate transplantation is not possible.

Slow freezing was discovered mainly by trial and error. To improve on the method, cryobiologists have tried to understand why it works. It's not easy to sort out the complex phenomena happening inside cells during cooling. One thing is clear: The formation of ice inside cells is typically lethal. It's not known for sure why intracellular ice kills, but cryobiologists know to avoid it at all costs. Slow freezing is intended to cool cells so that the water outside freezes to ice while the water inside does not.

How is that possible? If you live in a cold climate, you've probably seen people scattering salt on the sidewalk during a winter snow. This prevents ice from forming (and people from falling), because salt water freezes at a lower temperature than pure water. The higher the concentration of salt, the lower the freezing point. When cells are cooled slowly, water is gradually sucked out of them owing to a force known as osmotic pressure. The water remaining in the cell becomes saltier and saltier, and hence resists icing. If cells are cooled too rapidly, however, their contents don't become salty enough, and they freeze, with deadly consequences.

Slow freezing is not completely benign, because prolonged exposure to intracellular saltiness, though not as deadly as ice, is still damaging to cells. Furthermore, slow freezing does not prevent the formation of ice outside cells, which damages some tissues and all organs. To solve these problems, some researchers instead employ cooling methods that turn liquid water into an exotic state of matter that is said to be glassy or "vitrified," from the Latin word for glass. The vitrified state is solid but not crystalline. Its water molecules remain disorganized; they're not arranged into the orderly lattice you see in ice crystals.

Under normal circumstances, vitrification requires extremely rapid cooling, which is feasible for cells but not entire organs. Alternatively, you can get water to vitrify even at slow cooling rates if you add extremely high concentrations of cryoprotectants. Fertility researchers are already applying this method to oocytes and embryos, with great success.

Greg Fahy, who works at a company called 21st Century Medicine, has worked for decades to develop and refine various cocktails of cryoprotective agents for vitrifying organs. As his concoctions improved over the years, tissues looked healthier after chilling and rewarming, judging from various methods such as electron microscopy. But disappointingly, vitrified organs still failed the acid test: They didn't survive and function after rewarming and transplantation. In a remarkable advance, Fahy's team has at last succeeded, demonstrating recently that a previously vitrified kidney functioned after transplantation into a rabbit, lasting for weeks until the experiment was terminated. Inspired by Fahy's research, Alcor now uses vitrification to preserve the corpses of its members.

So how long can those corpses stay frozen without damage? You've probably noticed that items in your freezer do not last indefinitely. This has no bearing on cryonics, because the −196 degrees of liquid nitrogen is far colder than your freezer gets. It is closer to the lowest temperature possible — "absolute zero," or −273 degrees. Cold temperatures preserve because they slow down chemical reactions, the transformations that alter the atomic structure of molecules. The extreme cold of liquid nitrogen halts chemical reactions almost completely. The molecules in the corpses do not change, except when they are hit by cosmic rays or other types of ionizing radiation. Since such collisions are rare, the cryobiologist Peter Mazur has estimated that cells should last for thousands of years in liquid nitrogen. The clock may be ticking for Alcor members, but they have at least a few millennia before their time runs out.

There's a more fundamental problem, though. The Alcor members were all *dead* before they were vitrified, for minutes, hours, or sometimes even days. Isn't death irreversible, by definition? If so, how could reanimation ever succeed?

Irreversibility is indeed a central aspect of our definition of death. This makes the definition problematic. Irreversibility is not a timeless concept; it depends on currently available technology. What is irreversible today might become reversible in the future. For most of human history, a person was dead when respiration and heartbeat stopped. But now such changes are sometimes reversible. It is now possible to restore breathing, restart the heartbeat, or even transplant a healthy heart to replace a defective one.

Conversely, even if the heartbeat and respiration continue, a person with sufficiently severe brain damage is now regarded as legally dead. This redefinition was spurred by the introduction of mechanical ventilators in the 1960s. These kept accident victims alive so that the heart still pumped, even though the patient never regained consciousness. Eventually the heart stopped, or family members requested removal of the ventilator. At autopsy, the organs of the body looked perfectly normal to the naked eye or under a microscope. But the brain was discolored, soft or partially liquefied, and often disintegrated as it was removed. From this condition, nicknamed "respirator brain," pathologists concluded that the brain had died well before the rest of the body.

In the 1970s the United States and United Kingdom began to institute new laws governing the determination of death. To the traditional criterion of respiratory/circulatory failure, the United States added an alternative criterion: death of the entire brain, including the brainstem. In the United Kingdom, the death of the brainstem alone was considered sufficient. The U.S. definition is sometimes called "whole-brain death," while the U.K. one is known as "brainstem death."

The brainstem is critical for both respiration and consciousness. Its neurons generate signals that control the breathing muscles. If they fall inactive, breathing stops, and the patient cannot live without a mechanical ventilator. It is the brainstem's role in breathing that gives brainstem death its close tie to the traditional notion of respiratory/circulatory death. Another role played by the brainstem, perhaps even more important, is that it arouses the rest of the brain to consciousness. Our level of arousal goes up and down all the time, most dramatically in the sleep–wake cycle. Several populations of brainstem neurons, collectively called the reticular activating system, send their axons widely over the brain. These neurons secrete special neurotransmitters known as neuromodulators, chemicals that "wake up" the thalamus and cerebral cortex. Without them the patient cannot be conscious, even if the rest of the brain is intact.

The situation can be summarized this way: "If the brainstem is dead, then the brain is dead, and if the brain is dead, the person is dead." That's the rationale for the U.K. notion of brainstem death, and it makes sense because the brainstem typically functions longer than any other part of the brain. Damage to the brain causes cerebral edema, an abnormal buildup of fluid. This raises the pressure in the skull, causing blood flow to stagnate. Even more cells die, causing more edema and further shutting down the blood flow. The vicious cycle continues, and culminates with the brainstem being crushed by the pressure. So if the brainstem no longer functions, it's likely that the rest of the brain has already been destroyed.

This is the normal course of events. But sometimes — rarely — the entire brainstem is destroyed while the rest of the brain is left intact. The patient will never breathe without a mechanical ventilator, and will never regain consciousness. Yet one could argue that the patient still lives, assuming that memories, personality, and intelligence are preserved in the cerebrum. These properties seem more fundamental to personal identity than respiration, circulation, or brainstem function.

Today this distinction is merely theoretical, because no patient with complete brainstem damage has ever regained consciousness. But imagine a future medicine in which physicians can induce neurons in the brainstem to regenerate, reversing the damage. Then it might be possible for the patient to become conscious and functional again. The idea that the failure of the brainstem means that the person has died could eventually seem as outmoded as considering someone dead after respiratory/circulatory failure that is reversible.

Such future developments may seem far-fetched, but prognostication is not the real goal here. Rather, these thought experiments should motivate us to find a definition of death that is more fundamental. Ideally, the definition should remain valid no matter how far medicine progresses in the future. In this book I've talked about various ways of testing the hypothesis "You are your connectome." If this hypothesis is true, a fundamental definition of death follows immediately: Death is the destruction of the connectome. Of course, we don't know yet whether a connectome contains a person's memories, personality, or intellect. Testing these ideas will occupy neuroscientists for a very long time.

In the near term, all we can do is speculate. It's possible that a connectome contains most of the information in a person's memories. But even if that's the case, a connectome might not contain *all* of the information. Like any kind of summary, a connectome leaves out some details. Some of that discarded information could be relevant to personal identity. I conjecture that *connectome death* implies loss of a person's memories. However, the converse may not be true. Some of the information in a person's memories might be lost even if the connectome is perfectly preserved. (I'll tackle the issue of *completeness* in the next chapter.)

In its emphasis on brain *structure,* connectome death departs from conventional definitions based on brain function. The legal definition of death is the irreversible loss of *function* of the whole brain or of the brainstem. But as we've seen, the term *irreversible* is problematic. Snakebites and certain drugs can mimic brainstem death, but this loss of function is reversible. After mechanical ventilation for a short period, the patient recovers completely. So even for an expert, it can be tricky to decide when loss of function is permanent.

On the other hand, connectome death is based on a structural criterion that implies a truly irreversible loss of function (assuming that it implies the loss of memories). Alas, this definition is practically useless in a hospital. Currently, in live patients we can measure brain function through reflexes mediated by the brainstem, brain waves (EEG), or functional MRI. But we know of no way to find neuronal connectomes of living brains.

I can think of only one practical application of the idea of connectome death. Perhaps it's not really *that* practical, but I find it fascinating nonetheless. Why not use connectomics to critically examine the claims of cryonics? I've described at length the ways in which the brains of Alcor members have been damaged by circulatory/respiratory death and vitrification. Is there any chance that this damage could be reversed, as Alcor claims? To find out, I propose that we attempt to find the connectome of a vitrified brain. If the information in the connectome turns out to be erased, then we can declare connectome death. Resurrection by an advanced civilization of the future might be possible, but only for the body, not for the mind. If, however, the information is still intact, then we cannot rule out the possibility of resurrecting memories and restoring personal identity.

I suppose we should not conduct this experiment on a vitrified human brain. But Alcor has also vitrified the brains of some dogs and cats, at the request of pet-loving members. Perhaps some of these members would be willing to sacrifice their pets' brains in the name of science?

Until this scientific test is conducted, we can only speculate about what it might find. It's well-known that the brain is extremely sensitive to oxygen deprivation. Loss of consciousness follows in seconds, permanent brain damage after a few minutes. This is why disruption of blood flow to the brain can be so deadly, as happens in a stroke. At first glance, this seems like bad news for Alcor members. By the time Alcor receives the corpse, the brain has been deprived of oxygen for hours at least, and no living cells may remain. (Of course, it can be as difficult to define life and death for a cell as for the whole body.) Whether dead or alive, the cells have been badly damaged. Electron microscope (EM) studies have characterized the types of damage present in brain tissue a few hours after respiratory/circulatory death. Among other changes, mitochondria look damaged, and the DNA in the nucleus is abnormally clumped.

But these and other cellular abnormalities are irrelevant for connectome death. What matters is the integrity of synapses and "wires." Synapses seem less of a problem; they are still intact in the EM images, so they appear to be stable even in a dead brain. The status of axons and dendrites is harder to judge. Their cross-sections look largely intact in the published two-dimensional images, but there are some damaged locations. The big question is whether the damage has actually broken the "wires" of the brain. This can be answered by attempting to trace the neurites in three-dimensional images. If there are few breaks, tracing might still be possible. One could deal with an isolated break by bridging the gap between two free ends that were obviously once joined. But if there are clusters of many adjacent breaks, it might be impossible to figure out which free ends were once joined together. This would be true connectome death, a loss of information about connectivity that can never be recovered, no matter how advanced the technology.

At the present time, cryonics is closer to religion than to science, because it is based on faith rather than evidence. Its members believe that a future civilization will be able to resurrect them, based only on their faith in limitless technological progress. The test that I propose is a way of finally bringing some science to Ettinger's Wager. If the vitrified bodies contain intact connectomes, this does not prove that resurrection will be possible. But if connectome death has already occurred, resurrection will almost certainly be impossible.

Many Alcor members might not be eager to see the results of such a test. They may prefer blind belief as a means of consolation about their impending demise. If a scientific test has the potential to uncover factual information refuting their beliefs, they might prefer that the test not be conducted. There may be other members, though, who want evidence over faith, and would demand tests of connectome integrity.

It could turn out that the Alcor members stored in liquid nitrogen are already connectome dead. If so, that would not be the end of Alcor. They could always use connectomics as a means to improve their methods of preparing and vitrifying brains. Short of actually resurrecting their members, this is the only way I can imagine assessing the quality of their procedures. Even if their current method does not prevent connectome death, they could ultimately find one that does.

Cryonics is not the only way to preserve a body or a brain for the future. In his 1986 nanotechnology manifesto, *Engines of Creation,* Eric Drexler proposed that brains be preserved by chemical means. In a

1988 paper modestly titled "A Possible Cure for Death," Charles Olson independently proposed the same thing.

What Drexler and Olson were proposing was not a new procedure, but a new use for an old procedure called plastination. You may have seen one of the popular traveling exhibitions of human bodies preserved in plastic. Similar methods have long been used to prepare tissue for electron microscopy. The goal goes beyond merely preserving the look of tissue to the naked eye. Researchers try to leave every cellular detail intact, down to the structure of individual synapses. First, special chemicals like formaldehyde are delivered to cells by circulating them through the blood vessels. These are called fixatives, because they create links between the molecules that make up cells, fixing them in place. Once reinforced in this way, cellular structures are protected from disintegration. Then the water in the brain is replaced by alcohol, which in turn is replaced by an epoxy resin that hardens in an oven. The final product is a plastic block containing brain tissue (see Figure 53, left). The block is hard enough that it can be cut thinly with a diamond knife, as we do when finding connectomes.

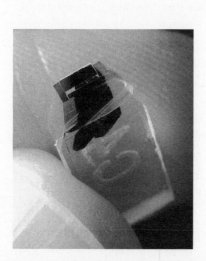 

FIGURE 53. *Plastination: brain tissue preserved in epoxy* (left) *and insect in amber* (right)

Aldehdye fixation, the first step of plastination, is also used by morticians when preserving bodies. This practice is called embalming, and is used to prepare bodies for temporary public display at funerals. In rare cases, the public display doesn't end with the funeral. For example, the Russian revolutionary Vladimir Lenin was embalmed after his death in 1924, and his body can still be seen in a Moscow mausoleum. It's not clear how long an embalmed body will remain intact. And even if it appears normal, its microscopic structure may be deteriorating. The full plastination procedure preserves biological structures indefinitely. The result looks similar to insects trapped in fossilized amber (Figure 53, right), some of which are millions of years old.

Plastination could be safer than cryonics, because it does not depend on a constant supply of liquid nitrogen. If Alcor goes bankrupt, or some kind of disaster damages its warehouse, the bodies and brains would be jeopardized. But a plastinated brain requires no special maintenance. Charles Olson predicted that "the cost of brain chemopreservation could be less than that of a typical funeral." There is an important stumbling block, though: Right now, plastination works on only very small pieces of brain. For various technical reasons, no one has yet succeeded in preserving an entire human brain with its connectome intact.

Ken Hayworth recently decided to do something about this. As you'll recall, he invented ATUM, the machine that slices brains thinly and collects them on a plastic tape for imaging and analysis. Many neuroscientists are driven not only by curiosity but also by ambition. Some want to discover something about the brain that will yield their next publication or promotion. Others aspire to win a Nobel Prize. But Hayworth makes all their ambitions look pedestrian. *His* goal is to live forever. As Woody Allen said, "I don't want to achieve immortality through my work. I want to achieve it through not dying."

Hayworth and his colleagues have established the Brain Preservation Prize, which offers $100,000 to any team that can successfully preserve a large brain in a way that leaves the connectome completely intact. A quarter of the prize money can be won by preserving a mouse brain. This is regarded as a steppingstone to a human brain, which is a thousand times larger in volume.

Hayworth is planning to plastinate his own brain. He would like to do this well *before* he dies of natural causes, while his brain is perfectly healthy. That would best preserve his brain for the future, but, by any ordinary definition, it would also kill him. He may have difficulty finding helpers, because their acts would likely be regarded as assisted suicide. Hayworth argues that plastinating his brain would not be suicide but salvation. It's his only chance at eternal life.

But how do you revive a plastinated brain? Raising the temperature brings cryopreserved sperm back to life. One can imagine thawing the bodies in the Alcor warehouse, but reversing aldehyde fixation and epoxy embedding seems much more difficult. Then again, if a civilization of the future is advanced enough to resurrect the dead, maybe they will also be advanced enough to unplastinate them. Eric Drexler imagined that an army of "nanobots," robots as tiny as molecules, might be used to unplastinate bodies and brains and repair whatever damage they've suffered. This dream certainly sounds attractive, but has not yet been realized.

Hayworth has thought carefully about his plans. If his plastinated brain cannot be revived, there might be an even better alternative. He imagines a future version of his ATUM invention, scaled up to handle a large brain — his brain. Once cut into ultrathin slices, his brain will be imaged and analyzed to find his connectome. The information will be used to create a computer simulation of Hayworth, one that thinks and feels like the real thing. This plan seems even more far-fetched than cryonics. Could it really be feasible?

FIFTEEN

Save As . . .

T'S DISTRESSING HOW little they tell us about heaven. We can at least imagine the gates. They are pearly and perched on a cloud. Saint Peter stands guard, ready to make sinners sweat by posing tough questions. But what is it like inside the gates? Everyone wears white. (I'm not sure how I feel about that.) The harp is the only accessory, and angels abound. These snippets of information aren't much to go on. Only recently did I realize why religions might prefer to be vague: People would rather fantasize about their own heaven than have one thrust upon them.

In the world's cultures and religions, conceptions of heaven have evolved slowly throughout history. Late in the second millennium, a radically new one emerged:

Heaven is a really *powerful computer.*

I don't mean that ecstatic look some nerds get when fondling their laptops. Let's not mistake such fetishism for a sign of spiritual enlightenment. But then again, why do these people spend so many of their waking hours online? Would it be too far-fetched to say that they thirst for transcendence, that they yearn to escape the inadequacies of this body and this world? While online, teenagers can forget the embarrassment of their pimply faces and underdeveloped physiques. People can take a pseudonym, alter their age, or masquerade with a photo of their dog. Netizens are free to be who they want to be, rather than who they really are.

A body chained to a computer, glassy eyes staring at a glowing screen, and fingers pecking away on a keyboard. That's a slightly less corporeal existence, to be sure, but I would only call it purgatory. It's still not what I mean by the new idea of heaven. Some nerds want more. They would like to discard their bodies completely and transfer their minds to computers. The idea of living as a computer simulation has been embraced by science fiction, which calls it "mind uploading," or "uploading" for short.

It's not possible yet, but perhaps all we have to do is wait for computers to get more powerful. Video games are stunning proof that computers can simulate the physical world. Every year the scenery looks more detailed and lush; every year bodies move in more lifelike ways. If computers can do that, why can't they simulate minds?

It's no exaggeration to compare uploading with ascension to heaven. Just think about the word itself. "*Up*loading" gets the direction right, as most agree that heaven is located in a high place. Some devotees prefer to say "mind downloading," but they are in the minority. It's not hard to understand why — "downloading" sounds suspiciously like going to hell.

Like thoughts of a traditional heaven, belief in uploading helps us cope with fear of death. Once uploaded, we would become immortal. But that's just the beginning. In the virtual world, we could beautify and strengthen our bodies simply by reprogramming the computer simulation. No need to suffer at the health club. Or perhaps we'll rise above such superficial concerns and focus instead on improving our minds. Let's not just upload — let's upgrade!

You may protest that uploading does not truly free us from the material world. The computer that runs the simulation might still malfunction or decay. But Christians also teach that the immortal soul does not lack a body in heaven. (Only during the time interval between death and Judgment Day does the soul wander bodiless.) There is still a body, but luckily it's incorruptible, an improved or perfected version.

Likewise, you'd be far better off living in a computer than in a body. Even if Alcor members turn out to be the lucky beneficiaries of bodily resurrection, enjoying the benefits of eternal youth provided by future medicine, they will have to worry about freak accidents that destroy their brains beyond repair. In contrast, the uploaded will feel safe and secure. They can always be restored from backup copies if they happen to be snuffed out by faulty hardware, or by a bug in whatever operating system of the future everyone will love to hate.

Some will no doubt say that all these arguments miss the point. Going to heaven isn't just about leaving one's bodily existence behind. It's about union with God. While uploaders might not get to meet the Christian God, they do expect to enter a new spiritual plane. Inside the great computer in the sky, the uploaded will mingle their lines of code to form a "hive mind," or collective consciousness. They will finally dissolve the distinction between self and other that lies at the root of evil and suffering, according to Buddhist teachings. With all the memories of humanity at its disposal, but none of the failings, this new superbeing will possess an unearthly wisdom that could be deemed godlike. We will find spiritual sustenance in our union with each other. Uploading will far surpass the Summer of Love and the Age of Aquarius, which bloomed briefly before the flower children grew up to drive BMWs and vote for lower taxes.

Enough about the advantages of uploading. Heaven sounds great. How can I get there? Well, that's a tougher question. As I'll explain in this chapter, only one even remotely plausible method has been proposed so far: simulating the electrical signals circulating in the network of neurons in your brain. A computer powerful enough to handle such a simulation could conceivably exist by the end of this century. To properly wire up the model neurons in the simulation, it would be necessary to find your connectome. Right now we can't envision any way of doing that without destroying your brain in the process. That sounds worrisome, but the Christian heaven thing isn't any better: Getting there almost always requires dying first. And there is an additional bonus to destructive uploading—it eliminates the troublesome question of what to do with your old self afterward.

For the sake of discussion, let's ignore these issues and simply suppose that your connectome could be found. Would that make uploading possible too? Simulating an entire brain is science fiction right now, but simulating *part* of a brain has been science since the 1950s at least. The models of perception, thought, and memory described in Part II have been formalized in mathematical equations and simulated on computers, although of course with goals less ambitious than uploading. The simulations are meant to reproduce a small subset of the brain's functions, as well as measurements of neural spiking from neuroscience experiments.

Carving, codebreaking, and comparing connectomes, as I envisioned them in Part IV, will depend on computers for analyzing large amounts of data, but they will not require the simulation of neural spiking. Having done some simulations myself, I regard this as a virtue. Analyzing data is less likely to lead us astray. Starting from the data, we extract what knowledge we can, with a minimum of assumptions. Simulation, in contrast, starts from the wish to reproduce an interesting phenomenon and tries to find the data necessary to do it. Wishful thinking can be dangerous if it's not based on reality. In the past, we've had to incorporate all kinds of assumptions into our models that are not backed up by empirical data. But connectomics and other methods of measuring from real brains are becoming more sophisticated. With better data, we'll be able to make our brain models more realistic. There's no denying that simulation will be a powerful way of doing neuroscience, when we can do it right.

Earlier I described how we might someday read a memory from a connectome by unscrambling its neurons to find a synaptic chain.

This would enable us to guess the order in which neurons spike during recall of a sequential memory. An alternative approach is to use the connectome to build a computer simulation of the spiking of neurons in a network, then run the simulation and watch the neurons to see the order in which they spike during memory recall. It's only natural to dream of scaling up this approach to an entire brain. Uploading is the ultimate way of testing the hypothesis "You are your connectome."

Researchers have engaged in protracted debates over the proper way to simulate the brain. The discussion of uploading in this chapter will raise all of the same conceptual difficulties, though — I hope — in more vivid form. Let's consider the first question that any modeler must answer: What constitutes success?

The promises of Alcor, resurrection and eternal youth, are easy to imagine. But uploading is a different story. What would it be like to live as a simulation inside a computer? Would you feel bored and lonely?

This question has been explored by the "brain in a vat" scenario, a staple of science fiction and college philosophy courses. Suppose that a mad scientist captures you, removes your brain, and manages to keep it alive and functioning in a vat of chemicals. Neural activity would still come and go, but would have no relation to the external world because of your brain's disembodiment. The isolation would far exceed lying in bed and closing your eyes. Severed from your sense organs and muscles, you would be enclosed in the darkest, most solitary confinement possible.

It's not a pretty picture, but uploaders need not worry. Any future civilization advanced enough to create a brain simulation would also be able to handle its input and output. Actually, input and output would be easy in comparison, because the connections between the brain and the external world are far less numerous than the connections *within* the brain. The optic nerve, which connects the eye to the brain, carries visual input through its million axons. That may sound like a lot, but there are many more axons running within the brain. (Most of the brain's 100 billion neurons have axons.) On the output side, the pyramidal tract carries signals from the motor cortex to the spinal cord, so that the brain can control movement of the body. Like the optic nerve, the pyramidal tract contains a million axons. Therefore, our future civilization could hook the simulation up to cameras and other sensors, or to an artificial body. If these "peripherals" were well crafted, the uploaded would be able to smell a rose and enjoy all the other pleasures of the real world.

But why stop at simulating the brain? Why not the world too? The uploaded could smell a virtual rose and pal around with other simulated brains. Many people seem to prefer virtual worlds these days anyway, judging from the time and money spent on computer games. And who knows? Maybe our physical world is actually a virtual world. If it were, would we have any way of knowing? Some physicists and philosophers — and those modern-day sages known as movie directors — suggest that we and the entire universe are actually simulations running on a gigantic computer. We may dismiss this idea as absurd, but logical reasoning cannot exclude it.

If the simulation feels exactly the same as reality, then living as a simulation will be just as much fun as real living. (Or for those who don't like the latter much, let's put it this way: Living as a simulation won't be any worse.) Audiophiles attempt to achieve "high fidelity" through electronic systems that faithfully reproduce a live musical performance. Uploaders will be obsessed by verisimilitude of a much more important kind. They can only hope for a very good approximation, not an exact replica. How accurate is accurate enough?

Most problems in computer science are straightforward to define. If we want to multiply two numbers, it's clear what success means. The goal of artificial intelligence (AI) is more difficult to state precisely. The mathematician Alan Turing provided an operational definition in 1950. He imagined a test in which an examiner interrogates a human and a machine. The examiner's task is to decide which is which. This might sound easy, but there is a catch: The interrogation is conducted by typing and reading text, in the style of Internet "chat." This prevents the examiner from distinguishing by appearance, sound, or other properties that Turing deemed irrelevant to intelligence. Now suppose that many examiners attempt the task. If this panel cannot come to the correct consensus, then we can declare the machine a successful example of AI.

Turing proposed his test to evaluate generic AI. We can easily refine it to measure success at simulating a specific person. Just restrict the examiners to friends and family, those who know the person best. If they are unable to distinguish between reality and simulation, then uploading has been successful.

Should sight and sound be barred from the specific Turing test, as they were from the generic version? You might balk at this, since voice and smile seem integral to the experience of loving someone. But people have fallen in love through Internet chat and email, before ever meeting each other. The surgical procedure of tracheotomy, which cuts a hole into the windpipe to relieve obstructed breathing, has the side effect of damaging the voice, yet everyone agrees that it's the same person afterward. A final reason to exclude the body from the test is that uploaders hope to escape their bodies. It's only their minds they care about preserving.

Will friends and family be vigilant enough to detect all differences between the simulation and the real person? Historical cases of impostors don't inspire confidence. In the sixteenth century, a man appeared in the French village of Artigat claiming to be Martin Guerre, who had been missing for eight years. He moved in with Guerre's wife and had children with her. Eventually accused of being an impostor, the "new" Guerre was acquitted at the first trial but found guilty at the second. He was on the verge of winning his appeal when another man dramatically appeared and claimed to be the real Guerre. All family members were suddenly unanimous in declaring the new Guerre — the man on trial — an impostor. He was convicted, and confessed to his crime shortly before his execution.

The new Guerre had excelled at imitation, failing only at side-by-side comparison. He might have survived a proper Turing test, conducted without sight or sound, as the real Guerre turned out not to remember his married life that well.

This and other cases of impostors show that friends and family are not perfect judges of personal identity. But if the differences are too subtle to be noticed, perhaps they don't matter. And even if they are noticeable, the simulation might not be considered a complete failure. Victims of brain damage are not the same after their injury, yet they are still accepted by others. If friends and family are the "customers" for uploading, their satisfaction is all that counts.

Then again, maybe the real customer is you, the person who wants to be uploaded. Of course it's important that your friends and family welcome the digitized you. But it's even more important that *you* be satisfied. This issue leads us onto shaky ground, but we can't avoid confronting it.

Suppose that you are uploaded to a computer. I turn the power switch on for the first time, and the simulation starts to run. I'm sure I would ask you, "How do you feel?" as if you were waking up from a deep sleep, or coming out of a coma. How would you reply?

The Turing test strives for objectivity by appealing to external examiners, but it would be silly to ignore subjective evaluation. Surely I'd want to ask your uploaded self, "Are you satisfied with your simulation?" We would never ask this of an equation that models a chemical reaction or a black hole, but it would be completely appropriate for a brain simulation.

At the same time, it's not clear whether I should believe your response. If your brain simulation malfunctions, you might act like a victim of brain damage. Neurologists know that such victims often deny their problems. Amnesics, for example, sometimes accuse others of deceiving them when they have memory lapses. Stroke victims don't always acknowledge paralysis, and may contrive fantastic explanations as to why they cannot perform certain tasks. Your subjective opinion simply might not be reliable.

Yet one could certainly argue that it's your opinion that should count the most. The satisfaction of your friends and family would depend on how well your simulation conforms to their expectations of your behavior. These expectations would be based on models of you, which they have constructed through years of observing your behavior. But you also have a self-model based on introspection as well as self-observation. Your self-model is based on far more data than someone else's model of you.

Perhaps there have been times when you've thought, "I'm not feeling like myself today." Maybe you've lost your temper over something trivial, or behaved in some other way that you found uncharacteristic. But usually you behave in a way that you expect. Your self-model would presumably be uploaded along with all your other memories. You would be able to check the fidelity of your simulation by continuously comparing your behavior with the predictions of your self-model. The more accurate the simulation, the fewer the inconsistencies.

Now let's suppose that uploading has been judged successful by both objective and subjective criteria. Your friends and family say they are satisfied. You (your simulation, that is) say you are satisfied. Can we now declare the uploading a success? There's one final catch: We do not have direct access to your feelings. Even if you say you feel fine, how do we know that you feel anything at all? Perhaps you're just going through the motions. What if uploading turned you into a zombie?

Some philosophers believe that it's fundamentally impossible to simulate consciousness on a computer. They say that a simulation of water, no matter how accurate, isn't actually wet. Similarly, your simulation might seem accurate to your friends and family, and might even proclaim its satisfaction, while still lacking the subjective experiences that we call consciousness. That may not seem bad, but it certainly doesn't sound like a route to immortality.

There is no way to refute the zombie idea, because there is no objective way to measure subjective feelings. In fact, the idea is so powerful that it can be applied to real brains as well as simulations. For all you know, your dog could be a zombie. It may act hungry, but it doesn't really have the feeling of hunger. (The French philosopher René Descartes argued that animals are zombies because they lack souls.) For all I know, you're a zombie too. There is no proof otherwise, because no person has direct experience of anyone else's feelings. Yet most people, especially pet lovers, believe that animals can feel pain. And virtually everyone believes that other humans feel pain.

I don't see any way to resolve such philosophical debates. It's just your intuition against mine. Personally, I think that a sufficiently accurate brain simulation would be conscious. The real difficulty is not philosophical but practical: Can that level of accuracy really be achieved?

Henry Markram has become famous as the creator of the world's most expensive brain simulation, but neuroscientists know him best for his pioneering experiments on synapses. Markram was one of the first to investigate the sequential version of Hebb's rule in a systematic way, by varying the time delay between the spiking of the two neurons when inducing synaptic plasticity. When I first heard Markram speak at a conference, I also encountered the chain-smoking and charming Alex Thomson, another prominent neuroscientist, who lectured about synapses with bubbling enthusiasm. She was in love with them, and wanted us to love them too. Markram, in contrast, came across as the high priest of synapses, summoning our awe and respect for their intricate mysteries.

In a 2009 lecture Markram promised a computer simulation of a human brain within ten years, a sound bite that traveled around the world. If you view the video of the lecture online, you might agree with me that his handsomely sculpted face looks a bit fierce, but his manner of speaking is gentle and inviting, with the quiet conviction of a visionary. He didn't sound so calm later that year. His competitor, the IBM researcher Dharmendra Modha, announced a simulation of a cat brain, after having claimed a mouse brain simulation in 2007. Markram responded with an angry letter to IBM's chief technology officer:

Dear Bernie,

You told me you would string this guy up by the toes the last time Mohda [sic] made his stupid statement about simulating the mouse's brain.

I thought that . . . journalists would be able to recognize that what
IBM reported is a scam — no where near a cat-scale brain simulation, but somehow they are totally deceived by these incredible statements.

I am absolutely shocked at this announcement. . . .

I suppose it is up to me to let the "cat out of the bag" about this
outright deception of the public.

Competition is great, but this is a disgrace and extremely harmful
to the field. Obviously Mohda would like to claim he simulated the
Human brain next — I really hope someone does some scientific and
ethical checking up on this guy.

All the best,

Henry

Markram didn't keep his indignation secret. He sent copies of the
letter to many reporters. One of them blogged about the controversy
with a story wittily headlined "Cat Fight Brews Over Cat Brain."

The letter marked a new low point in Markram's relationship with
IBM. They had started out allies in 2005, when IBM signed an agreement with Markram's institution, the École Polytechnique Fédérale
in Lausanne, Switzerland. The goal of the joint project was to show-
case IBM's Blue Gene/L, at that time the fastest supercomputer in the
world, by using it to simulate the brain. Markram called the project
"Blue Brain," an allusion to IBM's nickname, "Big Blue." But the relationship soured when Modha started a competing simulation project
at IBM's Almaden Research Center.

Markram tried to defend his own work by accusing his competitor
of fakery. But actually he cast doubt on the whole enterprise. Anyone
can simulate a huge number of equations and *claim* it's like a brain.
(You don't even need a supercomputer these days.) What's the proof?
How do we know that Markram isn't a scammer too?

His glitzy supercomputer should not distract us from a potentially
fatal flaw of his research: the lack of a well-defined criterion for judging
success. In the future, Blue Brain could be evaluated with the specific
Turing test explained earlier, but this test only becomes useful when

the simulation approaches the real thing. These purported mouse and cat brain simulations are not even in the ballpark yet. No "Mouse-tin Guerre" is going to fool you any time soon. The Turing test will tell us when we've reached our destination, but until that day comes, we need a way of knowing if we're going in the right direction.

Are these researchers really making progress? The full text of Markram's letter was too long to print here, so I'll just summarize the science behind his vitriol. In short, Blue Brain is composed of model neurons that are highly sophisticated in their handling of electrical and chemical signals. They are more faithful to real neurons than are the model neurons of Modha's simulation, which in turn are more realistic than the weighted voting model discussed in this book.

There is plenty of empirical evidence that the weighted voting model approximates many neurons well. But we also know that the model is not perfect, and can even fail badly for some neurons. Markram is correct that real neurons have many complexities that are not captured by simple models. A single neuron is an entire world in itself. Like any cell, it's a highly complex assembly of many molecules, a machine built from molecular parts. And each of these molecules in turn is a minuscule machine made of atoms.

As I mentioned earlier, *ion channels* are an important class of molecule, because they are responsible for the electrical signals in neurons. Axons, dendrites, and synapses contain different types of ion channels, or at least have them in differing numbers, which is why these parts of neurons have distinct electrical properties. In principle, every neuron is unique in its behavior, owing to the unique configuration of its ion channels. This is a far cry from the weighted voting model, according to which all neurons are essentially the same. But it sounds like bad news for brain simulation. If neurons were infinitely diverse, how could we ever succeed at modeling them? Measuring the properties of one neuron would tell you nothing about another.

There is one hope for escaping the morass of infinite variation: neuron types. You may recall that Cajal classified neurons into types based on location and shape. You can think of these properties as being like an animal's habitat and appearance. When a neuroscientist speaks of the double bouquet cell of the neocortex, it reminds me of the way that a naturalist speaks of the polar bear of the Arctic. The naturalist might also point out that polar bears, unlike brown bears, all hunt for seals. Likewise, neurons of the same type generally exhibit the same electrical behaviors. This is presumably because their ion channels are distributed in the same way.

If this is the case, then neural diversity is actually finite. We should compile a catalog of all neuron types, a "parts list" for the brain, and then construct a model for each type. We'll assume that each model is valid for all neurons of that type in all normal brains, much as we assume that all resistors behave the same way in any electronic device. Once all neuron types have been modeled, we'll be ready to simulate brains.

Markram's laboratory has characterized the electrical properties of many neocortical neuron types through experiments in vitro. Based on this data, they have modeled each neuron type as hundreds of interacting electrical "compartments," which is an approximation to simulating the millions of ion channels in a neuron. Markram deserves credit for the realism of the multicompartmental model neurons used in Blue Brain.

But Blue Brain is severely lacking in one respect. Since no cortical connectome is known yet, it's not clear how to connect the model neurons with each other. Markram follows Peters' Rule, a theoretical principle stating that connectivity is random. The accidental collisions of axons and dendrites in the tangled "spaghetti" of the brain lead to contact points. At every one of these, a synapse occurs with some probability, as if it were the outcome of tossing a biased coin.

Peters' Rule is conceptually related to an idea introduced earlier, the random synapse creation of neural Darwinism. The ideas are not equivalent, however. Neural Darwinism includes activity-dependent synapse elimination, which makes the surviving connections end up nonrandom. Violations of Peters' Rule have already been discovered. I suspect that many more will be found, and that the rule has managed to survive only because of our ignorance of connectomes.

As computer scientists like to say, "Garbage in, garbage out." If the neural connectivity of Blue Brain is wrong, the simulation will be too. But let's not be overly critical. In the future, Markram could always incorporate information from connectomes into Blue Brain. Then wouldn't his simulation become truly realistic?

To answer this question, let's again consider the roundworm *C. elegans*. Its connectome is already known, unlike that of the neocortex. It may come as a surprise that only small parts of its nervous system have been simulated. These models have been helpful for understanding some simple behaviors, but they are piecemeal efforts. No one has come close to simulating the entire nervous system.

Unfortunately, we lack good models of *C. elegans* neurons. As I mentioned earlier, most of them don't even spike, so the weighted voting model isn't valid. To model the neurons, we'd have to measure from them, but this turns out to be more difficult for *C. elegans* than for mouse or even human neurons. We also lack information about *C. elegans* synapses. The connectome did not even specify whether the synapses were excitatory or inhibitory.

So Blue Brain lacks a connectome, while *C. elegans* lacks models of neuron types. Both elements are needed to simulate a brain or nervous system. Thus the earlier claim should be revised to say, "You are your connectome plus models of neuron types." (Let's assume that a connectome is defined to specify the type of each neuron.) But the models of neuron types are likely to contain much less information than the connectome, as most scientists agree that there are far fewer neuron types than neurons. In this sense, "You are your connectome" would remain a very good approximation. Furthermore, we assumed above that all neurons of one type behave in the same way in all normal brains, just as all polar bears hunt seals under normal circumstances. If we uploaded multiple people, all the simulations could share the same models of neuron types. The only information unique to a person would be his or her connectome.

It's worth noting that the balance of information content is quite different in *C. elegans*. Its three hundred neurons have been classified into about one hundred types, which is not that much smaller than the number of neurons. Essentially every neuron (along with its twin on the other side of the body) is its own type. If every neuron ends up requiring its own model, the total information in all the models might exceed that in the connectome. So "You are your connectome" would be a terrible approximation for a worm, even though it might be almost perfect for us.

To put it another way, the *C. elegans* nervous system is like a machine built from parts that are all unique. The individual workings of the parts are just as important as their organization. The opposite extreme would be a machine built from a single type of part. (You may be old enough to remember old-fashioned Lego sets, which contained only one type of Lego block.) The functionality of such a machine would depend almost entirely on the organization of its parts.

Electronic devices are close to this extreme, as they contain only a few types of parts, like resistors, capacitors, and transistors. That's why a radio's wiring diagram determines so much of its function. The parts list for the human brain is longer, so it will take many years of effort to model every neuron type in the human brain. But the parts list is still far shorter than the total number of parts. That's why the organization of the parts is so important, and why "You are your connectome" may turn out to be a very good approximation.

There's one more important aspect of connectomes to include in brain simulations: change. Without it, your uploaded self would not be able to store new memories or learn new skills. Markram and Modha have included reweighting using mathematical models of Hebbian synaptic plasticity. But it's also important to include reconnection, rewiring, and regeneration. In general, our models for the four R's are much less refined than those for electrical signals in neurons. It will be possible to improve them, but it will take many more years of research.

These are all important caveats, but models of neuron types and connectome change still fit into the overall framework of connectome-based brain simulation. Is there anything about the brain that is fundamentally incompatible with the framework? One difficulty is that neurons can interact outside the confines of synapses. For example, neurotransmitter molecules might escape from one synapse, and diffuse away to be sensed by a more distant neuron. This could lead to interactions between neurons not connected by a synapse, or even between neurons that do not actually contact each other. Because this interaction is extrasynaptic, it is not encompassed in the connectome. It might be possible to model some extrasynaptic interactions fairly simply. But it's also possible that the diffusion of neurotransmitter molecules in the cramped and tortuous spaces between neurons would require complex models.

If extrasynaptic interactions turn out to be critical for brain function, then it might be necessary to reject the hypothesis "You are your connectome." The weaker statement "You are your brain" could still be defensible, but this would be much more difficult to use as a basis for uploading. We might have to throw away the abstraction of the connectome and descend still further to the atomic level. One could imagine using the laws of physics to create a computer simulation of every atom in a brain. This would be extremely faithful to reality, much more than a connectome-based simulation.

The catch is that a huge number of equations would be necessary, since there are so many atoms. It seems absurd to even consider the enormous computational power required, and is completely out of the question unless your remote descendants survive for galactic time scales. At the present time, it's difficult to simulate even those modest assemblies of atoms called molecules. Simulating all the atoms of a brain is almost beyond imagining. Limited computational power is not the only barrier. There is also the difficulty of obtaining the information to initialize the simulation. It might be necessary to measure all the positions and velocities of the atoms in the brain, which is far more information than in a connectome. It's not clear how to collect that information, or how to do it in a reasonable amount of time.

So if you're an uploader, your only hope is a connectome-based strategy. Over the coming years, we'll find out whether "You are your connectome" is true or at least a good approximation, through the types of research discussed in Part IV. Such scientific research will be focused on more near-term goals, but it will also give us some idea of the chances that uploading will actually work.

As humans, we have long believed — or wanted to believe — that there is more to life than material existence: "I'm more than a piece of meat. I have a soul." As a dream about escaping the body, uploading is no more than the latest iteration of an enduring wish.

Over the past few centuries, science has shaken our belief in the soul. First we were told, "You are a bunch of atoms." According to this doctrine of materialism, the universe is a gigantic pool table, and atoms are like billiard balls moving and colliding according to the laws of physics. Your atoms are no exception to this rule, and obey the same laws as all the other atoms in the universe. Then biology and neuroscience told us, "You are a machine." According to this doctrine of mechanism, the parts of your machine are cells or special molecules like DNA. Your body and brain are not fundamentally different from the artificial machines manufactured by humans, only much more complex.

But computers have forced us to reexamine the doctrines of materialism and mechanism. "You are a bunch of information," uploaders believe. You are neither machine nor matter. Those are just means of storing what you really are — information. In our everyday experiences with computers, we have learned to distinguish between information and its material incarnation. Suppose I take your laptop computer and, overcome by a murderous rage, hack it to pieces. You retrieve its carcass and manage to pull out its hard drive, which is still in good shape. You don't have to mourn long. Just transfer the information to another laptop, and we can go about our lives as if nothing ever happened.

Uploaders don't see a fundamental difference between humans and laptops. They think it should be possible to transfer the information of your personal identity into some other material form. The uploader chides the materialist by saying "You are not your atoms, but the pattern in which they are arranged." The uploader rebukes the mechanist with "You are not your neurons, but the pattern in which they are connected." Although a pattern requires matter for embodiment, it belongs to the abstract world of information, not the concrete world of matter.

Indeed, the uploader might say that your new laptop is the *reincarnation* of your old laptop. The transmigration of your laptop's soul occurred when you transferred the information in the hard drive. And so we are led to the idea that *information is the new soul*. We've come full circle, returning to the idea that the self is based on a nonmaterial entity, something that is ghostlier than matter.

The analogy is not perfect. Unlike the soul, which is usually regarded as immortal, information can be lost permanently. The nanotechnologist Ralph Merkle has defined the concept of *information theoretic death* as the destruction of the information about personal identity stored in the brain. Returning to our laptop example to illustrate his idea, suppose that the original hard drive from your damaged computer is recovered, but that its motor was damaged during the rampage. It's beyond your technical capabilities to transfer the information to your laptop. But someone with superior nerdly powers might be able to fix the motor so that you can perform the transfer. On the other hand, if I were really mean, I could have passed a powerful magnet over your hard drive, instead of chopping up your computer. This would have erased the information on the hard drive, which is stored in a magnetic pattern. In that case, no technology, no matter how advanced, could recover your information. It's fundamentally impossible.

Merkle's definition of death is of more philosophical than practical importance. To apply it, we need to know exactly how memories, personality, and other aspects of personal identity are stored in the brain. If this information is contained in the connectome, then information theoretic death is nothing more than connectome death.

All efforts to achieve immortality can be viewed as attempts to preserve information. Most humans would like to have children before they die. Some of the information in their DNA will survive in their children's DNA, and other kinds of information will survive in their children's memory. Some humans try to achieve immortality by writing songs or books that will be remembered by future generations. This is yet another attempt to embed information about themselves in the minds of others.

Cryonics and uploading seek to preserve the information in brains. They can be viewed as part of a broader movement called transhumanism, which seeks to transform the human species. We no longer have to wait for the glacial course of Darwinian evolution, say the transhumanists; we can use technology to alter our bodies and brains. Or we can discard them completely, and migrate to computers.

Transhumanism has been ridiculed as the "rapture of the nerds." Some find it strange to fantasize about eternal life in the future when so many dire problems threaten the world today. But transhumanism is the inevitable and logical extension of Enlightenment thought, which exalted the power of human reason. Emboldened by their successes in mathematics and science, European thinkers sought to establish law and philosophy on principles deduced from rational thought, rather than appealing to tradition or revelation from God. The philosopher Leibniz even believed that all disagreements arose from mistakes in reasoning, and suggested that they could be resolved by formalizing arguments with symbolic logic.

But in the twentieth century the limitations of reason became painfully apparent. The logician Kurt Gödel proved that mathematics is incomplete, because there exist statements that are true but cannot be proved. The physicists who pioneered quantum mechanics discovered that some events are truly random and cannot be predicted even with infinite information and computational power. If reason fails even in mathematics and science, how can we expect it to succeed elsewhere? Indeed, many philosophers have become convinced that morality cannot be derived from reason; they call attempts to do so the "naturalistic fallacy."

Transhumanists no longer believe that reason can answer all questions. Yet they still believe in its supremacy, because of its power to continually create more advanced technologies. Transhumanism resolves a major problem of the Enlightenment, which was based on a scientific worldview that deprived many people of the feeling of purpose. If physical reality is just a bunch of atoms bouncing around, or genes competing to replicate, then life may seem meaningless. In his book on the Big Bang, *The First Three Minutes*, the theoretical physicist Steven Weinberg wrote, "The more the universe seems comprehensible, the more it also seems pointless." This viewpoint was expressed more poetically by Pascal in his *Pensées*:

> I see those frightful spaces of the universe which surround me, and I find myself tied to one corner of this vast expanse, without knowing why I am put in this place rather than in another, nor why the short time which is given me to live is assigned to me at this point rather than at another of the whole eternity which was before me or which shall come after me. I see nothing but infinites on all sides, which surround me as an atom and as a shadow which endures only for an instant and returns no more. All I know is that I must soon die, but what I know least is this very death which I cannot escape.

The "meaning of life" includes both universal and personal dimensions. We can ask both "Are we here for a reason?" and "Am *I* here for a reason?" Transhumanism answers these questions as follows. First, it's the destiny of humankind to transcend the human condition. This is not merely what will happen, but what should happen. Second, it can be a personal goal to sign up for Alcor, dream about uploading, or use technology to otherwise improve oneself. In both of these ways, transhumanism lends meaning to lives that were robbed of it by science.

The Bible said that God made man in his own image. The German philosopher Ludwig Feuerbach said that man made God in his own image. The transhumanists say that humanity will make itself into God.

EPILOGUE

IT'S TIME TO return to reality. We've each got one life to live, and one brain to do it with. In the end, every important goal in life boils down to changing our brains. We are blessed with natural mechanisms for transformation, but we find their limitations frustrating. Beyond appealing to our curiosity and sense of wonder, can neuroscience give us new insights and techniques for changing ourselves?

I've argued that one of the most important ideas of our time is connectionism, the doctrine that emphasizes the importance of connections for mental function. According to this notion, changing our brains is really about changing our connectomes. Connectionism dates back to the nineteenth century, but empirical evaluation of its claims has been difficult. At long last, thanks to the emerging technologies of connectomics, we are poised to test the doctrine. Is it indeed true that minds differ because connectomes differ? If we succeed in answering that question, we will also be able to identify desirable changes in the brain's wiring.

The next step will be to devise new methods of promoting such changes, based on molecular interventions that promote the four R's: reweighting, reconnection, rewiring, and regeneration. The methods would also utilize training regimens that harness the four R's to bring about positive changes.

To realize all these advances, we must continue to develop the necessary technologies. In the history of science, there are many examples of conceptual barriers that could not be surmounted by researchers, however brilliant they were, until the right tools became available. You wouldn't expect a caveman to figure out the workings of an old-fashioned mechanical clock if he didn't have a screwdriver. In the same vein, it's unrealistic to expect neuroscientists to figure out the brain without extremely sophisticated tools. Our technologies are starting to become equal to the task, but we will need to make them many times more powerful.

We need to create a research environment that fosters these technological advances. One possibility is to undertake "grand challenges," ambitious projects that stimulate our imagination and mobilize our intellectual efforts. We could set a goal of finding the entire neuronal connectome of a mouse brain using electron microscopy, or the entire regional connectome of a human brain with light microscopy. The projects are of comparable difficulty, because they require the acquisition and analysis of similar amounts of data. I estimate that either would require a decade of intense effort. Both connectomes would be invaluable resources for neuroscientists, just as genomes have become indispensable to biologists.

These projects would be enormously difficult, but we could simultaneously pursue shortcuts. With the technologies developed, it would be possible to rapidly and cheaply find smaller connectomes. Compared with the grand challenges above, it should be a thousand times faster to find the neuronal connectome of a cubic millimeter of brain, or the regional connectome of a mouse brain. Finding many smaller connectomes would be important for studying individual differences and change.

Why should we invest in future technologies when we need to find better treatments for mental disorders right now? I think we should do both. Our therapies will surely improve over the next few years, but I expect that it will take decades to find true cures. Since this will be a continuing battle, it's worth making a reasonable investment today to reap rewards in the long run.

You may be skeptical that technology will ever progress enough to find connectomes quickly and cheaply. Before the Human Genome Project began, sequencing an entire human genome seemed almost impossible too. Connectomics might look difficult, but there's a certain sense in which it's trivial compared with the larger endeavor of neuroscience. Since the goal is well defined, we know exactly what success means, and can quantify progress. In contrast, the broader goal of neuroscience — to understand how the brain works — is only hazily defined. Even the experts don't agree about what it means. Once a goal is clearly defined, time, money, and effort are likely to yield progress. That's why I believe that connectomics will achieve its goals, however ambitious they might seem. We just need to rise to the challenge.

The young boy laughed as he splashed in the water. Returning to land, he asked, "Teacher, why does the stream flow?" The old man gazed silently at the novice and replied, "Earth tells water how to move." During their journey back to the temple, they crossed a precarious footbridge. The novice clutched the old man's hand tightly. He looked at the stream far below and asked, "Teacher, why is the canyon so deep?" As they reached the safety of the other side, the old man replied, "Water tells earth how to move."

I believe the stream inside our brain works in much the same way. The flow of neural activity through our connectomes drives our experiences of the present and leaves behind impressions that become our memories of the past. Connectomics marks a turning point in human history. As we evolved from apelike ancestors on the African savannah, what distinguished us was our larger brains. We have used our brains to fashion technologies that have given us ever more amazing capabilities. Eventually these technologies will become so powerful that we will use them to know ourselves — and to change ourselves for the better.

ACKNOWLEDGMENTS

David van Essen planted the seed for this book by inviting me to lecture at the 2007 meeting of the Society for Neuroscience. Speaking before an audience of thousands, I concluded by laying out the challenge of finding connectomes. Upon hearing the buzz that followed, Bob Prior encouraged me to write a book. I took his suggestion but decided to target the general public. Since no knowledge could be assumed, I would have to argue from first principles and question all my beliefs. I was following the prescription "Empty your cup so that it may be filled."

When I finished a draft in 2009, Catharine Carlin pointed me to Jim Levine, and Dan Ariely made the introduction. Jim's enthusiastic offer to serve as my agent was an enormous boost. He recruited the brilliant Amanda Cook, who has repeatedly prodded me with the question "Why should we care?" Beyond editing my writing and improving my storytelling, she has shaped my thinking. I never anticipated how drastically the book would change under her guidance, and I consider myself lucky that it did.

A life in science comes with a wonderful fringe benefit — opportunities to meet smart and interesting colleagues. Many fascinating discussions with other neuroscientists have enriched this book. The wise counsel of David Tank originally set me on the road to connectomes. The encouragement of Winfried Denk, who critiqued two drafts of the book, kept me writing. Jeff Lichtman patiently educated me about synapse elimination and neural Darwinism. Ken Hayworth explained his cutting machines and passionately argued the case for transhumanism. Daniel Berger contributed many suggestions for improving the book.

I am grateful to Scott Emmons and David Hall for information on *C. elegans,* Axel Borst on the fly brain, Kevin O'Hara on the California redwood, Misha Tsodyks and Haim Sompolinsky on associative memory models, Eric Knudsen and Stephen Smith on reconnection and rewiring, Carlos Lois and Fatih Yanik on regeneration, Mitya Chklovskii and Alex Koulakov on wiring economy, Kristen Harris on serial electron microscopy, Guyeon Wei on semiconductor electronics, Dick Masland and Josh Sanes on neuron types, Kathy Rockland and Almut Schüez on cortical anatomy, Harvey Karten and Jerry Schneider on brain evolution, Michale Fee on birdsong, Li-Huei Tsai and Pavel Osten on brain disorders, Vamsi Mootha on biology, Niko Schiff on neurology, Drazen and Danica Prelec on philosophy and psychology, and Michael Häusser and Arnd Roth on dendritic biophysics.

Mike Suh and John Shon assisted me with the initial proposal for the book. Along with Janet Choi and Julia Kuhl, they also commented on the final version. Scott Heftler suggested some fun comparisons. Fellow authors Sue Corkin, Mike Gazzaniga, Allan Hobson, and Lisa Randall advised me at critical junctures. The meticulous editing and impeccable logic of Katya Rice polished the prose, to my delight.

Several public speaking experiences attuned me to the zeitgeist. Ute Meta Bauer invited me to lecture for the Visual Arts Program at MIT, Susan Hockfield brought me to the World Economic Forum, and Sarah Caddick helped me spread the word through a 2010 TED talk.

Finally, I owe thanks to the Gatsby Charitable Foundation, the Howard Hughes Medical Institute, and the Human Frontiers Science Program for funding my research in connectomics.

NOTES

an estimated 10,000 synapses per neuron. This is likely an overestimate, and its exact value should not be taken too seriously. A more reliable enumeration has been performed for a brain structure called the neocortex, and yielded 0.16 quadrillion synapses (Tang et al. 2001).

xvii *and phallic graffiti everywhere:* Beard 2008.

xviii *notions of the self:* I'm indebted to Ken Hayworth for clarifying this point to me.

xix *contains 100 billion neurons:* A recent study places the average number at 86 billion (Azevedo et al. 2009).

1. Genius and Madness

3 *Ivan Turgenev:* The brains of Turgenev and other famous Russians are described in Vein and Maat-Schieman 2008.

Sir Arthur Keith: Keith 1927. Unfortunately for Keith and his reputation, he is remembered less for his scientific discoveries than for his endorsement of the Piltdown Man. These skull fragments, purported to be a "missing link" in the evolution of man from ape, were eventually exposed as fake. Piltdown Man became one of the most famous hoaxes in the history of science.

4 *French theoretical physicist:* Keith resolved his conundrum in a similar way, writing that "a detailed study of Anatole France's life, so far as it is known, shows us that he was in many senses a primitive man." He wrapped up his essay by reaffirming his belief that brain size and intelligence are actually related: "In the long run I expect it will be found that there is a close correspondence between brain mass and the degree of function subserved by that organ."

5 *average head size:* Galton 1889.

6 *people with bigger brains:* McDaniel 2005.

with high accuracy: If the correlation coefficient of two variables is r, then knowing one variable reduces the typical prediction error of the other by a factor of $\sqrt{1 - r^2}$.

correlation between IQ and brain volume: McDaniel 2005.

"Beauty Map": Galton recounts the story in the last chapter of his memoirs, about "Race Improvement, or Eugenics" (Galton 1908). In a three-volume hagiography, Karl Pearson reminisced about his mentor: "Galton, influenced by his own motto . . . , seldom went for a walk or attended a meeting or lecture without counting something. If it was not yawns or fidgets, it was the colour of hair, of eyes, or of skins" (Pearson 1924, p. 340). Galton.org pays tribute to the man.

7 *imbecile:* Pearson 1906. While Pearson confirmed Galton's finding that head size and school grades were statistically related, he also noted that head size was a poor predictor of school grades for any particular individual. Even handwriting quality was a better predictor than head size.

cerebrum, the cerebellum, and the brainstem: Swanson 2000 divides the brain more finely into the cerebral cortex, basal ganglia, thalamus, hypothalamus, tectum, tegmentum, cerebellum, pons, and medulla. Swanson argues that all of the many proposed schemes for coarsely dividing the brain can be regarded as different groupings of these nine basic parts. For example, in the tripartite scheme of Figure 7, the cerebrum is defined as the cortex plus the basal ganglia, and the brainstem as the rest of the parts minus the cerebellum. A book-length exposition of his views can be found in Swanson 2012. Note that some authorities exclude the thalamus and hypothalamus from the brainstem, so its definition is ambiguous.

spares mental abilities: Although introductory textbooks usually don't mention it, cerebellar damage does have some effects on emotion and cognition (Strick, Dum, and Fiez 2009; Schmahmann 2010).

8 *largest of the three parts:* The cerebrum is largest by volume, but the cerebellum has the most neurons, with an estimated 70 billion (Azevedo et al. 2009) or 100 billion (Andersen, Korbo, and Pakkenberg 1992). Almost all of these are the so-called granule cells. Because these are very small, the cerebellum takes up only 10 percent of the brain's volume (Rilling and Insel 1998). The neocortex, the dominant part of the cerebrum, is estimated to contain 20 billion neurons (Pakkenberg and Gundersen 1997).

into four lobes: The borders of the occipital lobe are defined with additional landmarks but are somewhat arbitrary. The four lobes are named for the four bones of the skull that overlie them. Some authorities define a fifth, limbic lobe. This is visible on the faces of the hemispheres exposed by cutting the cerebrum in half along the longitudinal fissure. Buried inside the Sylvian fissure is a part of the cortex known as the insula, which is large enough that some regard it as another lobe.

10 *prisons and mental asylums:* Micale 1985.

11 *not confining them in chains:* Harris 2003.

Figure 10: The lesion is centered in the inferior frontal gyrus (fold) of the left cerebral hemisphere. The story of the patient Leborgne, nicknamed Tan, is told in Finger 2005 and Schiller 1963, 1992.

12 *hemispheres looked so similar:* Researchers have also found slight structural asymmetries between the right and left hemispheres, but it's been difficult to tell whether these have anything to do with lateralization of function (Keller et al. 2009).

dominant for language: Rasmussen and Milner 1977. In a minority of left-handers and ambidextrous people, the right hemisphere is dominant for language, or both hemispheres are involved.

13 *Harvey sent specimens:* Abraham 2002; Paterniti 2000.

Sandra Witelson: Witelson, Kigar, and Harvey 1999.

brains of luminaries: Burrell 2004.

14 *his 1819 treatise:* Gall 1835.

IQ is correlated: Jung and Haier 2007.

15 *London taxi drivers:* Maguire et al. 2000.

In musicians: Hutchinson et al. 2003; Gaser and Schlaug 2003. My statement "thicker cortex" is a bit glib, because the measurements rely on a method called voxel-based morphometry, which can't distinguish between thickening and other structural changes. Thickening is just one possible interpretation.

Bilinguals: Mechelli et al. 2004.

severe mental disorder: Kessler et al. 2005.

17 *symptoms of autism:* Frith 2008.

18 *unable to function:* There are also milder forms of autism, which involve some but not all of the symptoms. For example, Asperger's syndrome is defined by social impairment and repetitive behaviors but not linguistic difficulties. The term *autism spectrum disorders* has been introduced to include the entire range, from mild to severe forms of autism. Fombonne 2009 estimated the incidence of full-blown autism as two per 1,000 people, and that of the autism spectrum disorders as several times higher.

"beautiful child": Frith 1993.

19 *defined the syndrome:* The Viennese pediatrician Hans Asperger is also credited with having defined autism a few years earlier.

large heads: Kanner 1943.

heads and brains: Redcay and Courchesne 2005. Interestingly, autism provides counterevidence to the maxim that bigger is better. Phrenologists might respond by pointing to autistic "savants," who exhibit impressive displays of memory, numerical calculation, or other mental abilities like the fictional character in the movie *Rain Man* (Treffert 2009). Perhaps these enhanced mental abilities could be explained by the enlargement of autistic brains. But most autistic children are not savants, and even savants have disabilities. Perhaps it's fairer to conclude that the phrenological approach of studying brain size is an oversimplification.

frontal lobe: Carper et al. 2002.

20 *first-person account:* BGW 2002.

less effective for the negative symptoms: Second-generation, or "atypical," antipsychotic drugs were marketed as superior for negative symptoms, but this claim is now being questioned. For more on this controversy, see Murphy et al. 2006 and Leucht et al. 2009. The atypicals are less likely to produce movement disorders as side effects, which were common with first-generation, or "typical," antipsychotics.

overall brain volume: Steen et al. 2006; Vita et al. 2006. The difference exists even in patients receiving their first psychiatric treatment, so it does not appear to be a long-term effect of antipsychotic medications.

lateral and third ventricles: Steen et al. 2006.

21 *"graveyard":* Plum 1972.

2. Border Disputes

23 *infant brain grows rapidly:* Voigt and Pakkenberg 1983.

philosophy of education: Spurzheim was actually quite sophisticated for his time, acknowledging that other changes in the brain might take place besides growth: "The growth of the organs [brain regions], however, is not the only or even most important advantage to be derived from proper exercise. . . . [T]he size of the organ . . . will not augment in proportion to its being exercised, but its fibres will act with more facility" (Spurzheim 1833, pp. 131–132).

through simple mazes: The Hebb–Williams test of animal intelligence was a battery of twenty-four problems, each involving finding food in a simple maze. Donald Hebb pioneered this type of research on the effects of environmental enrichment. It's briefly noted in Hebb 1949, which is better known for its presentation of Hebb's theories of the cell assembly and synaptic plasticity (see Chapters 4 and 5 below).

Mark Rosenzweig: Rosenzweig 1996. The test of statistical significance was based on comparisons of siblings born in the same litter. The change in cortical size was not due to an overall change in brain size. In fact, the noncortical areas of the brain were slightly smaller. The change was not due to an increase in body size either. The environmentally enriched rats were actually somewhat lighter, owing to their increased activity.

24 *learning to juggle balls:* Draganski et al. 2004; Boyke et al. 2008.

intensive study for exams: Draganski et al. 2006.

26 *Korbinian Brodmann:* His map spanned the neocortex, which is the predominant part of the cerebral cortex. Confusingly, the term *cortex* often serves as an abbreviation for the neocortex alone. Brodmann divided the cortex into forty-three areas (Brodmann 1909), but not all are visible in Figure 11, which includes only one view of the cerebrum. If you look closely, you'll notice that 52 is the map's largest number, not 43. That's because Brodmann skipped 12–16 and 48–51. He reserved these numbers for cortical areas in animals that appeared to have no analogues in the human cortex. Brodmann used a microscope to delineate the areas, as I'll describe in Chapter 10. However, the areas line up roughly with the cortical folds, so they can be located approximately even without a microscope.

after three months: Cramer 2008.

27 *after stroke:* Cramer 2008.

removing one hemisphere: Mathern 2010. The procedure is justified, for example, when MRI reveals a one-sided brain abnormality that is clearly the cause of the seizures.

walk and even run: Vining et al. 1997. For inspiring testimonials by patients, see http://hemifoundation.intuitwebsites.com.

migrate to the right hemisphere: Basser 1962 discusses very early childhood; Boatman et al. 1999, late childhood. The phenomenon was already noted by Broca in the nineteenth century.

28 *Miguel Nicolelis:* Nicolelis 2007.

crude act of butchery: Bagwell 2005. By medieval times the church had taken over the practice of medicine. A 1215 papal edict forbade the clergy from practicing surgery, because contact with blood or bodily fluids was considered contaminating. Surgery was left to barbers, who may have been more effective healers than the university-trained physicians.

29 *tie off large arteries:* Finger and Hustwit 2003.

unremarked for so long: The history of phantom limbs from Paré to Mitchell is surveyed in Finger and Hustwit 2003.

phantom is not real: Reilly and Sirigu 2008.

irritated nerve endings: This explanation is credited to Descartes by Finger and Hustwit 2003.

this didn't help: Ramachandran and Blakeslee 1999.

Wilder Penfield: Penfield and Boldrey 1937.

30 *V. S. Ramachandran:* Ramachandran, Stewart, and Rogers-Ramachandran 1992. An entertaining and readable account of this research is provided in Ramachandran and Blakeslee 1999. Ramachandran's discoveries in humans were probably not surprising to Mike Merzenich and other neuroscientists who had already found similar phenomena in animals, as reviewed in Buonomano and Merzenich 1998.

sensation of a phantom limb: This description may sound incomplete, because I've spoken only of functions and avoided inputs and pathways, which are discussed later in this book. It's more revealing to say that amputation deprives the lower-arm territory of inputs from sensory pathways. Remapping replaces these with sensory inputs from the face and upper arm.

31 *stroked the face of an amputee:* There was even a one-to-one mapping between facial locations and digits of the phantom hand (cheek to thumb, chin to pinkie, and so on).

Functional MRI: More precisely, fMRI measures the BOLD (blood oxygen level dependent) signal, which was discovered by the Japanese scientist Seiji Ogawa. This is defined as the ratio between the oxygenated and deoxygenated forms of hemoglobin, the molecule in the blood that ferries oxygen from the lungs to the rest of the body. Using a brain region has two opposing effects on the BOLD signal. First, the region burns more energy, which deoxygenates hemoglobin. Second, blood flow increases, which carries in more oxygenated hemoglobin. (Many believe that blood flow increases in response to use, because the brain precisely regulates blood flow to fulfill the energy needs of each region.) Since either of these effects can dominate, using a brain region can either increase or decrease the BOLD signal, which confuses the interpretation of fMRI. On a related note, since the BOLD signal reflects energy consumption, some people quip that using fMRI to understand the brain is like trying to understand the engine of a car by measuring where it gets the hottest.

"spots on brains": These images give the misleading impression that a person uses a small fraction of the brain for any given task. However, each image is actually obtained by subtracting two images corresponding to two similar mental tasks. A "lit-up" region was used *more* in one task than the other. One should not conclude that all the other regions lay idle. Many of them were active, but the level of activity was similar in both tasks.

32 *the shift occurred:* Lotze et al. 2001 also demonstrated a similar remapping of area 4 in

amputees, and measured brain activity caused by imagined movements of the phantom hand. Researchers also used fMRI to demonstrate remapping of area 4 in stroke patients. The hand representation moved up or down within area 4, depending on the location of brain damage. Further studies found that stroke can cause remapping on a larger scale, affecting distant areas on the same or the other side of the brain (Cramer 2008).

left-hand representation: Elbert et al. 1995 used magnetic source imaging rather than fMRI. They found a shift in the average location of the left-hand representation within area 3, which they interpreted as a change in area. But a direct measurement of the size of the representation showed no statistically significant change. They couldn't prove that the shift was caused by musical training, because of the possibility of selection bias. However, the size of the shift was correlated with the age at which musical training began. See Amunts et al. 1997 for a related study using MRI.

crippling disorders: Elbert and Rockstroh 2004.

focal dystonias: A famous example is the pianist Leon Fleisher, who lost the use of his right hand for thirty-five years but recently made a comeback with both hands after receiving treatment based on injections of Botox into his arm muscles.

violin and Braille: Sterr et al. 1998 not only showed an expanded hand representation but argued that the arrangement of the fingers in the representation was disorganized, which might distinguish Braille reading from violin playing.

frontal lobe in schizophrenics: Glahn et al. 2005.

33   *about brain disorders:* Kaiser et al. 2010 and Bosl et al. 2011 are two recent studies characterizing activity in the autistic brain.

strength with a machine: Actually the scientific studies use isometric measurements, meaning that the force is measured while the joint angle is held fixed. This is more controlled, because force depends on joint angle. Muscle size is quantified by cross-sectional area (CSA), which is expected to be roughly proportional to the number of fibers and hence to strength.

correlation coefficients: You might think it's silly to research this correlation, since common sense tells us it must be strong. Actually this has been surprisingly difficult to establish empirically. Maughan, Watson, and Weir 1983 reported lower correlation coefficients and took the contrarian view that "strength is not a useful predictive index of muscle cross-sectional area." More recent studies like Bamman et al. 2000 and Fukunaga et al. 2001 appear to agree on stronger correlations, possibly thanks to improvements in measurement methods. Still, many interesting questions remain unanswered. For example, is the relationship between size and strength different for powerlifters and bodybuilders, or for elite athletes and regular people?

34   *boundaries of Brodmann's map:* Lashley and Clark 1946.

cortical equipotentiality: Lashley 1929.

35   *over 100 million:* The estimate that Brodmann area 17 contains over 100 million neurons is from Huttenlocher 1990.

3. No Neuron Is an Island

40   *Figure 13:* Although in the brain no neuron is an island, isolated neurons can be artificially cultured in a plastic dish, as shown in Figure 13. Even this neuron is not truly island-like, though, as its branches actually extend far outside the borders of the image, to form connections with other neurons in the dish. The image was obtained by scanning electron microscopy.

41   *one* million: If we don't restrict ourselves to the brain, neurites can be longer still. Some neurites travel from the brain to the spinal cord, and others connect the spinal cord to the toes and fingers. And let's not forget that giraffes and whales have neurites too.

42 *marked "ax" and "sp":* "ax" marks an axon, and "sp" a dendritic spine, which sticks out of
the dendrite like a thorn.
do not really touch: Invisible in the image in Figure 14 are various molecules that span the
cleft between the membranes of the two neurons and bring them into direct contact. But
the whole notion of "touching" starts to break down at even higher magnification. What
we call matter consists mainly of empty space between its constituent particles.

44 *same small set of neurotransmitters:* Eccles et al. 1954 stated the principle that a neuron
secretes a *single* neurotransmitter, and attributed it to Sir Henry Dale, who won a 1936
Nobel Prize for his studies of synaptic transmission. Eccles 1976 later revised Dale's Principle to allow for multiple neurotransmitters. Eccles himself shared a 1963 Nobel Prize for
his work on synapses. More recently, researchers have found a further exception: neurons
are capable of switching from one neurotransmitter to another.

45 *brain secretes thoughts:* The eighteenth-century French philosopher and physiologist
Pierre Cabanis wrote that "the brain secretes thought as the liver secretes bile."
send them to specific targets: In most biological contexts, chemical signaling relies upon
the specificity of molecular binding (the lock-and-key mechanism). That's not sufficient
to prevent crosstalk between synapses, because many synapses use exactly the same neurotransmitter.

46 *minimize crosstalk:* That's not to say there is *zero* crosstalk. Some spillover of neurotransmitter is known to occur, and appears important for brain function in certain cases.

47 *"most expensive loveseat":* Russell 1978.
67 miles of tangled wire: Kolodzey 1981.
insulating material: Small amounts of crosstalk can still occur because of electrical fields,
which penetrate the insulation.
millions of miles: The brain is over a million cubic millimeters in volume, and a large fraction of that is cortex. According to Braitenberg and Schüz 1998, a cubic millimeter of cortex contains several miles of neurites.

48 *single axon, long and thin:* This description holds for a very common type of neuron,
the pyramidal neuron of the cortex. However, there are many other types of neurons,
which have different appearances. The dendrite–axon distinction is not even valid for
some types of neuron, especially in invertebrate nervous systems. For these neuron types,
each neurite both sends and receives synapses.
typical synapse is from: But there are also synapses from axon to cell body, dendrite to
dendrite, axon to axon, and pretty much any other variation you can think of.
Figure 17: This figure shows a brief segment of the voltage signal recorded from a neuron
in the hippocampus of a rat exploring a maze. The experiment is described in Epsztein,
Brecht, and Lee 2011.

49 *above the static:* After the telegraph, the telephone was invented for analog communication — that is, the transmission of voice signals without encoding them into pulses. But
now the telephone system has become digital again, utilizing something like Morse code.
The encoding and decoding are invisible to the user because they are done quickly and
automatically by electronic circuits rather than human operators. Why would our sophisticated telephone systems return to the style of communication used in the primitive
telegraph? One reason is that today's systems are designed to transmit information at the
highest possible rate. This requires operating at the limits set by noise, so the best strategy
is again digital.
spike triggers secretion: I say "passing" because synapses mostly occur at locations along
the axon, so that spikes propagate past them. Some synapses are located at axonal dead
ends, however, so that spikes terminate at them.
a synapse converts: How receptors transform chemical signals into electrical ones will be
explained in Chapter 6.

50   *toward the cell body:* This is known as the Law of Dynamic Polarization. Neuroscientists
      sometimes violate the law by using electrical stimulation to initiate a spike that travels
      backward along the axon toward the cell body. Such "antidromic" propagation is opposite
      the normal direction, proving that signal transmission along the axon is two-way.

51   *cells that support them:* The nervous system also contains non-neuronal cells, known as
      glia. These come in a number of types, and are absolutely essential for keeping the brain
      alive and functioning. I will take the traditional view that glial cells are like the crew, sup-
      porting the cast of neurons that star in the mental show. Neurons and glial cells are about
      equally numerous (Azevedo et al. 2009). Much more about glia can be found in Fields
      2009.

52   *synapses onto muscle fibers:* These are called neuromuscular junctions, to contrast them
      with ordinary synapses between neurons.
      "To move things": Sherrington 1924.
      190 stations: Bradley 1920.

53   *synapses are weak:* Some contrarians believe that there are a small number of strong syn-
      apses, and these are the important ones for brain function.
      cannot typically relay a spike: Even if synapses are weak, it's possible for a single neuron
      to drive another neuron to spike. The neurons need only be connected by a large number
      of synapses. However, this situation is apparently rare in practice.
      synapses made by the axon: Actually, synapses behave stochastically. With every spike,
      some randomly fail to secrete neurotransmitter.

54   *all possible pathways:* For the snake, your eye communicates with your legs and not your
      salivary glands. For the steak, it's the other way around. In telecom networks, such selec-
      tivity is achieved through the operation of *routing.* Every message has an address, which
      is separate from the content of the message. This is most obvious when you mail a let-
      ter. You write the address on the outside of an envelope, the content on the paper within.
      Similarly, you enter the address of a telephone by punching in its number to request a
      call, but it's the ensuing conversation that contains the content. A node in the network
      routes an incoming message by looking at its address and relaying it to another node that
      is closer to the destination specified by the address. A message takes a pathway through
      the network determined by these routing decisions. These are made by human workers
      in the post office, and by devices called switches in the telephone network. Even if a single
      pathway could relay spikes, it's not obvious how the nervous system could route spikes
      through the right pathway to reach a specific destination. Axons aren't doing any rout-
      ing; they just send spikes indiscriminately to all their synapses. Perhaps routing could be
      found elsewhere in the neuron, but there is a fundamental problem with the whole idea.
      Since a spike is merely a pulse, it's unclear how it could carry both the content and the
      address of a message. This is why telecom networks are probably not such a good meta-
      phor for the brain. That being said, this theoretical argument cannot exclude the possibil-
      ity that messages consist of sequences of spikes, that assemblies of neurons can function
      as routing devices, and that the brain is like a communication network when examined
      at a higher level of organization. In fact, some theorists still contend that the routing op-
      eration is helpful for understanding brain function (Olshausen, Anderson, and Van Essen
      1993).
      If dendrites lack spikes: As explained in Häusser et al. 2000 and Stuart et al. 2007, re-
      searchers have challenged the traditional conception that dendrites don't spike. Experi-
      ments on neurons kept alive in slices of brain have demonstrated spikes in dendrites. If
      this phenomenon also occurs in intact brains, it could be that each dendrite of a neu-
      ron takes a vote of its synapses, and then the cell body takes a vote of its dendrites. This
      would be analogous to the American presidential election, in which the people of each
      state vote in the general election, and then the states vote in the Electoral College. In

principle, it's possible for a candidate to win this two-stage election without winning the popular vote.

55 *quantifies the weight:* This is a simplification, as the notion of the "strength" of a synapse is more complex than can be summarized in a single number.

"weighted voting model": Engineers call this the "linear threshold model" of a neuron, to contrast the summation in voting, which they call a "linear" operation, with thresholding, *a "nonlinear" operation:* Yet another name for the model is "simple perceptron."

ranging from milliseconds: This is yet another dimension in which chemical synapses are more versatile than electrical synapses.

56 Inhibitory *synapses:* More-direct evidence for the importance of synaptic inhibition comes from studies of movement. Muscles are generally organized in pairs with opposing effects. The biceps and triceps muscles, which are on either side of your upper arm, are one example. The biceps bends your elbow; the triceps extends it. Your nervous system is constantly sending spikes to both the biceps and the triceps. This is why your muscles are not completely relaxed at rest; they have some degree of "muscle tone." When you bend your elbow, your nervous system sends more spikes to your biceps, causing it to contract, and simultaneously sends fewer spikes to your triceps, causing it to relax. One reason for this reduction is that the motor neurons controlling the triceps receive inhibition from synapses.

tends to "inhibit" spiking: In a more accurate definition, excitatory versus inhibitory depends on whether the so-called reversal potential for the synapse is above or below the threshold voltage at which a neuron spikes.

another kind of synapse: An electrical synapse, or gap junction, consists of a cluster of molecules, each of which is a tiny tunnel connecting the interior of one neuron to the interior of the other.

other limitations: Electrical synapses are less versatile in many other ways. The duration of synaptic currents is fixed and short. Electrical current generally flows in both directions, though it may flow more readily in one of them. If two-way sounds superior to one-way, you might regard electrical synapses as more powerful than chemical synapses. But two-way communication between neurons can be established by two chemical synapses, one in each direction, while electrical synapses cannot establish one-way communication. Therefore two-way communication is actually a limitation. Electrical synapses are known to play an important role when a population of neurons needs to generate spikes simultaneously. Fast bidirectional communication makes sense for achieving such synchronicity. Electrical synapses exert only electrical effects, while chemical synapses can additionally trigger molecular signals within the receiving neuron. The extra steps in chemical transmission may slow it down, but they also allow for amplification, and modulation by other processes.

how should our voting model be revised: A simpler effect of inhibition on pathways almost goes without mentioning: A single pathway containing a mixture of inhibitory and excitatory synapses can't relay spikes, however strong the synapses may be.

veto many excitatory synapses: In 1943, the theoretical neuroscientists Warren McCulloch and Walter Pitts introduced the first voting model of a neuron. The McCulloch–Pitts model adhered to the slogan "One synapse, one vote," but only for excitatory synapses. An inhibitory synapse was allowed to have complete veto power over many excitatory synapses. It can be shown that the McCulloch–Pitts model is a special case of the weighted voting model, just by giving the inhibitory synapse a very large weight.

makes only excitatory synapses: This follows from Dale's Principle, because a given neurotransmitter generally has the same electrical effect on any neuron, either always excitatory or always inhibitory. (The sign of the electrical current depends on the molecular machinery on the receiving side of the synaptic cleft.)

A similar uniformity: Also, the uniformity does not extend to strength; a neuron can make a strong synapse onto one neuron and a weak synapse onto another.

57 *most neurons are excitatory:* The split is 80–20 in the cortex.

58 *increases its selectivity:* Here's another way of thinking about the significance of selective spiking. Nature has gone to the trouble of preventing crosstalk between wires. Why do this when signals are mixed at every neuron by convergence and divergence? The answer is that selectivity is preserved because neurons often fail to spike.

albeit a very different kind: As computers have pervaded our everyday lives, we have lost sight of how strange they really are. A digital computer is a machine like no other, because of its universality. Like an infinitely versatile Swiss Army knife, a computer can perform any kind of computation if equipped with the right software. (This is an informal statement of the Church–Turing thesis, which is formulated for an abstract computing model known as a universal Turing machine. It's something like a modern digital computer with a hard disk of infinite capacity.) This is very different from your toolbox, which contains a hammer, a screwdriver, a saw, a wrench, and a drill, all of which are specialized for different functions. Since brain regions are specialized for particular functions, the brain is more like your toolbox than like a universal computer. Just as the structures of a saw and a hammer are closely related to their functions in carpentry, the structures of brain regions are likely to be closely related to their functions.

59 *deviate somewhat from the voting model:* The weighted voting model is only an approximation to a real neuron, which may be more complex. Bullock et al. 2005 briefly describes inaccuracies of the approximation, and Yuste 2010 is a book-length review of the properties of dendrites.

4. Neurons All the Way Down

61 *make scientific observations:* Quiroga et al. 2005.

62 *photo of Julia Roberts:* Fried's experiment was striking because it was done in humans. His results are less surprising if you're familiar with the work of his predecessors, who did similar experiments in monkeys and other animals. For example, Desimone et al. 1984 reported neurons that responded selectively to faces.

celebrity supercouple: Actually there were a few spikes, though not many. Fried and his colleagues did find another group of neurons in the same person that was selectively (dare I say nostalgically?) activated by Aniston and Pitt together, but not by Aniston alone.

"celebrity neuron": In a famous paper Horace Barlow called this the "grandmother cell" theory of perception, joking that there is a neuron in his brain that is active if and only if his grandmother is present (Barlow 1972). Gross 2002, however, credits the "grandmother cell" theory to Jerome Lettvin.

63 *small percentage:* This "small percentage" model actually fits the data better than the "one and only one" model. Before, I emphasized the neurons that responded to a single celebrity, but these were actually a small minority. Many more neurons responded to no celebrities in the experiment, and even fewer neurons responded to two celebrities. To see that this is consistent with the "small percentage" model, compare the random sampling of celebrities with throwing darts while blindfolded. Finding a celebrity that activates a neuron is like hitting the dartboard; both events have low probability. It's most likely that no dart will hit the dartboard. If you're lucky, one dart will make it. It's very unlikely that two or more darts will. That being said, the experiment cannot rule out the existence of neurons that truly respond to just one celebrity. To identify such neurons, it would be necessary to show patients a huge number of photos.

number of possible patterns: Here we've simplistically defined the activity pattern to be binary: Every neuron is either active or inactive. We could refine the definition to include the rates at which the active neurons spike. Then the activity pattern would contain even more information.

Leibniz was wrong: The philosophically sophisticated may disagree with my claim, saying that Leibniz was referring not to perception but to qualia, the subjective feelings that accompany perception. In other words, he was really referring to consciousness, and measurements of spiking haven't told us much about that.

This kind of mind reading: Can fMRI also be used for mind reading? Recently some researchers have argued that fMRI could be used to detect when a person is lying (Langleben et al. 2002; Kozel et al. 2005). The standard "lie detector" used in criminal prosecution and employment interviews is the polygraph. This measures blood pressure, pulse, respiration, and skin conductivity, which are supposed to reveal the hidden emotional stress that usually accompanies the act of lying. There is widespread skepticism, however, about the accuracy of the polygraph, and because fMRI directly assesses mental state by measuring the activation of the brain, it could potentially be more accurate. In laboratory experiments, some researchers have claimed good results with using a brain scanner to distinguish between lying and truth-telling human subjects. Based on this research, businessmen have founded two new companies seeking to commercialize fMRI lie detection. It's still not clear whether fMRI will turn out to be superior to the polygraph, but that's irrelevant to the discussion here. The point is that fMRI researchers are hoping only for the crudest kind of mind reading. None of them would dream of using fMRI to read out a highly specific mental property like the perception of Jennifer Aniston.

64 *"the shoulders of giants":* Recently some revisionist historians have interpreted this remark as sarcasm rather than modesty, as it comes from a letter to rival scientist Robert Hooke, who was a hunchback. Newton and Hooke later became enemies because of a dispute over optics.

66 *"receives excitatory synapses":* You may have noticed something missing from this rule: inhibitory neurons. Most cortical neurons are excitatory, but we should not neglect the inhibitory neurons, as they surely have some function too. Recall that the "Jennifer Aniston neuron" did not spike for photos of Jen with Brad Pitt. We can emulate this behavior by adding to our construction an inhibitory synapse from a neuron that detects Brad. If this synapse is strong enough, then its vote will override the votes from the neurons that detect components of Jen, and keep the neuron silent if Brad is present. More generally, it has been theorized that inhibitory synapses are helpful for making fine distinctions between similar stimuli. Excitatory synapses may enable a neuron to spike for a certain type of nose, while inhibitory synapses enable it to *not* spike for similar types of noses.

hierarchical organization: Actually the part–whole rule was used to wire up only every other layer of his network. The other half were wired by another rule: A neuron receives excitatory synapses from neurons that detect slightly different versions of the same stimulus. The neuron has a low threshold for spiking and therefore responds to any of the stimulus variations. This rule is required for achieving another important property of perception: its invariance to "irrelevant" differences between stimuli.

perceptron: Some use *perceptron* to refer only to the case of a single layer of synapses, and specify *multilayer perceptron* for the more general case. But Rosenblatt originally meant the term to refer to a multilayer network, and I follow his usage here.

the layer just below: The perceptron has a feature that is not consistent with the known connectivity of the brain. Its pathways go only from the bottom of the hierarchy to the top. In real brains, there are also connections going in the opposite direction. What could be the role of these top-down pathways in perception, and how are they likely to be organized? In the "interactive activation" model of McClelland and Rumelhart 1981, a letter-detecting neuron receives bottom-up connections from neurons that detect the strokes of the letter. (Such part-to-whole connections were discussed in the main text.) But this fails to explain a simple phenomenon: How do you know that the middle letter of *C–T* is likely to be *A*, *O*, or *U*, and not *E* or *I*? In the interactive activation model, a letter-detecting neuron also receives top-down connections from neurons that detect words containing the letter. In the above example, an *A* detector is assumed to receive a connection from a *CAT* detector. More generally, one can imagine the rule "A neuron that detects a whole sends excitatory synapses to neurons that detect its parts." This allows a neuron to detect a stimulus by weighing evidence received from *both* bottom-up and top-down connections.

68 *people who have blue eyes:* It's because many wholes can share a single part that a hierarchical representation is more efficient than a flat one.

connectionism: The term *connectionism* more commonly refers to a 1980s movement in cognitive science that sought to explain the human mind using model networks of weighted voting neurons. Philosophers of mind argued over its merits relative to the "symbolic" approach of understanding the mind as a digital computer. As this heated debate recedes into history, it's better to use the word in the broader sense I've defined, as an intellectual tradition that dates back to the nineteenth century and is still evolving.

69 *perception or thought:* The MTL is regarded by some as the top of the hierarchy hypothesized earlier (see Figure 51). At the bottom are areas of the cortex devoted to perception alone. Thinking does not activate the neurons in these areas, or at least not so much. The dividing line between perception and thinking does not appear to be sharp. Rather, the involvement of neurons in thinking appears graded, increasing gradually as one ascends the hierarchy.

72 *never function perfectly:* According to some theorists, inhibitory neurons may be more precise at controlling the spread of activity than neuron thresholds, providing for superior memory recall.

information overload: Inhibitory neurons increase memory capacity by retarding the spread of activity. To serve this dampening function, the connections of the inhibitory neurons don't need much organization at all. If each receives synapses from a random selection of excitatory neurons, it will be activated whenever the "mob" is active. If it sends synapses back to another random selection of excitatory neurons, it will exert a dampening effect on the crowd. An engineer would say that inhibitory neurons exert "negative feedback" on excitatory neurons. The household thermostat is the classic example of negative feedback. If the temperature of a heated room increases beyond a certain point, the thermostat turns off the heat; if the temperature decreases, the thermostat turns on the heat. In both cases the thermostat acts to oppose the change in temperature, in the same way that inhibitory neurons act to oppose changes in the activity of excitatory neurons. In this view, inhibitory neurons play a supporting role in brain function, so their connections don't have to be very specific.

73 *left to right:* Note that this looks like the perceptron shown earlier, but turned on its side. Although a synaptic chain can be viewed as a special case of a perceptron, it's quite different from the typical perceptron, which is used to model perception. The neurons in one layer of a perceptron typically detect different stimuli, so each is wired to a different subset of neurons in the previous layer. (Or if they are wired to the same neurons, the strengths of the synapses differ.) All the neurons in one layer of a synaptic chain get activated together, so their connections with the previous layer need not be different. The synaptic chain has been formalized in mathematical models by a number of researchers (see, for instance, Amari 1972 and Abeles 1982). The American theoretical physicist John Hopfield developed related models in the 1980s.

theory of connectionism: Donald Hebb proposed and named the cell assembly (Hebb 1949). Early computer simulations of model networks with cell assemblies were performed in the 1950s. The English theorist David Marr and the Japanese theorist Shun-ichi

Amari were two prominent researchers who studied the equations of such models using pencil and paper in the 1960s and 1970s (see, for example, Marr 1971 and Amari 1972). But the real heyday of connectionism came in the 1980s, following the seminal papers of John Hopfield (Hopfield 1982; Hopfield and Tank 1986). Using esoteric mathematical techniques from a branch of physics known as spin glass theory, theoretical physicists had a field day calculating memory capacity through a statistical treatment of the effects of overlap between cell assemblies (see Amit 1989; Mezard, Parisi, and Virasoro 1987; and Amit et al. 1985). By the time this flurry of activity petered out in the 1990s, these researchers had discovered many interesting properties of the models. Also around this time, the PDP Research Group, a collective of cognitive scientists, published an influential two-volume manifesto containing many interesting connectionist models (Rumelhart and McClelland 1986).

74 *"Problem of Serial Order":* Lashley attributed the "associative chain model" to the British psychologist Edward Titchener, citing a book from 1909. Actually both authors spoke of chains of psychological associations rather than neural connections. Strangely, Lashley did not use the word *synapse* in his article, although he was a neuroscientist. Nevertheless, the notion of synaptic chains is implicit in his writing.

huge variety of activity: There would also have to be points where two chains converge into one, or we would quickly run out of neurons.

problem of syntax: In a similar vein of criticism, some computer scientists have argued that relations between ideas are richer than simple associations. To say that the ideas of fish and water are associated does not do justice to their relationship. It's more richly descriptive to say that a fish "lives in" water. Computer scientists represent such relationships with a "semantic network," which looks like a connectome except that each arrow is labeled with a type of relation.

addressed Lashley's second: These connectionist models achieve greater computational power by introducing latent or hidden variables, to augment the variables that are used to represent explicit ideas.

5. The Assembly of Memories

76 *two-and-a-half-ton:* The blocks varied in size; this number is an estimate of the average (Petrie 1883). Most blocks were limestone, but some were granite.

2.3 million: Petrie 1883.

one hundred thousand workers: Herodotus wrote that one hundred thousand slaves labored for twenty years to transport the blocks from a distant quarry to the pyramid. Many recent Egyptologists have disagreed, arguing that the main quarry was nearby, the workforce was far smaller, and the workers were not slaves.

77 *"There exists in the mind":* Plato, *Theaetetus.*

straight-edged instrument: Draaisma 2000.

Artisans and engineers: The term *plasticity* comes from materials science. A *plastic* material holds its new shape when deformed; an *elastic* material bounces back to its original shape. Because wax is plastic, it can hold an impression and hence store information about the past. In this technical usage, *plastic* is an adjective that refers to the behavior of materials in response to deformation. *Plastic* is more commonly used as a noun, to refer to any of the synthetic polymer materials widely used in manufactured products. The common usage is related to the technical usage in that these materials can undergo plastic deformations at higher temperatures, a feature that is often used in manufacturing. These materials are usually elastic at room temperature, however. Furthermore, there are other types of materials, such as metals, that can also undergo plastic deformations.

78 *a phenomenon I'll call* reconnection: The reweighting–reconnection distinction is clean-

est when there is at most one synapse from neuron A to neuron B, as is assumed generally in this chapter. The distinction is blurred if there are multiple synapses from A to B. Then synapse creation and elimination might leave the neurons connected, and only change the number of synapses that A sends to B. This would alter the weight of A's vote in B's spiking, bringing about reweighting rather than reconnection.

In the 1960s: My claim of "most neuroscientists" comes from hearsay and is difficult to document rigorously. One example is the Australian neuroscientist Sir John Eccles, who wrote that learning involves "growth just of bigger and better synapses that are already there, not growth of new connections" (Eccles 1965). Rosenzweig 1996 provides some historical review from the viewpoint of a neuroscientist, but the issue should be examined by a real historian.

Figure 23: This image is based on data from an experiment described in Yang, Pan, and Gan 2009.

79 *created and eliminated:* Spines have also been observed to change in size, which suggests that synapses are changing in strength.

80 *counting synapses:* Greenough, Black, and Wallace 1987.

neo-phrenological theory: Some researchers looked at the sizes of synapses as well as their numbers. There is evidence that the size of a synapse is correlated with its strength. The researchers found that enriched environments increased the average size of synapses in the rat cortex. However, one should not equate learning with an increase in synaptic size, just as one should not equate it with an increase in synaptic number. Other experiments have demonstrated decreases in the average size of synapses. Which of these changes dominates depends on the particular location in the cortex as well as the layer of the neurons involved.

82 *Hebb:* Hebb 1949. The sequential rule was also proposed in the late nineteenth century by the Scottish philosopher Alexander Bain (see Wilkes and Wade 1997), but his theory never took hold. Perhaps Bain had the misfortune of living too early, when so little was known about the brain. He knew about fibers and pathways, and guessed the existence of connections between pathways, but the existence of neurons or synapses had not yet been established.

Hebbian plasticity refers: Less is known about plasticity of synapses involving inhibitory neurons, so that will not be discussed here. According to the conventional wisdom, the connections between excitatory neurons are more specific and more shaped by learning. Those involving inhibitory neurons are relatively indiscriminate and may be less influenced by learning.

isolate the spikes: This method, known as "single-unit" recording, was pioneered by the English scientist Edgar Adrian, who garnered the 1932 Nobel Prize and eventually the title "Lord."

83 *mouth of a speaker:* Synapses onto muscles had already been studied in the 1930s and 1940s. In the 1950s Sir John Eccles and other researchers refined the method of intracellular recording and applied it to synapses in the spinal cord. Eccles went on to share a 1963 Nobel Prize for his efforts.

by injecting electrical current: The text describes using two intracellular electrodes to study a specific pair of neurons. This is the most precise method of studying synapses, and is relatively recent. Eccles used a single intracellular electrode to record from one postsynaptic neuron, and stimulated a large number of presynaptic neurons by injecting current through an extracellular wire.

The size of this blip: If there happen to be multiple synapses from neuron A to neuron B, then the size of the blip is the aggregate strength of all the synapses.

Repeated stimulation: Bi and Poo 1998; Markram et al. 1997. Whether sequential or simultaneous stimulation is more effective at inducing plasticity depends on the type of neurons involved. Strictly speaking, these experiments did not demonstrate change in single synapses. There were multiple synapses between the measured pair of neurons, and the experiments demonstrated a change in their aggregate strength. In general, such experiments have trouble distinguishing between reweighting and reconnection. If the interaction between two neurons strengthens, it could be the result of an increase in the number of synapses between them, not just synaptic strengthening. Another interesting issue, which I don't have space to discuss here, is the mechanism by which a synapse detects simultaneous or sequential spiking. This appears to happen through a special molecule called the NMDA receptor.

84 *Brad and Jen:* It's simplistic to say that you forgot the relationship between Jen and Brad. Although they're no longer married, you still remember that they *used* to be married. To represent this knowledge, you could imagine that there is a marriage neuron and a divorce neuron. Initially the cell assembly includes the Brad, Jen, and marriage neurons. Later on, the cell assembly includes the Brad, Jen, and divorce neurons. This solution is still not entirely satisfactory, but a better solution would have to confront Lashley's critique that connectionism cannot represent syntax, and is outside the scope of this book.

two neurons are weakened: For example, Stent 1973 proposed that the synapse from A to B is weakened if A is repeatedly inactive while B is active. Other variants were proposed by many theorists. The flip side of the sequential version of Hebb's rule is: If two neurons are repeatedly activated sequentially, the connection from the second to the first is weakened. Empirical evidence was found by Markram et al. 1997 and Bi and Poo 1998. In combination with Hebb's rule, it's known as "spike-timing dependent plasticity."

85 *direct competition:* Miller 1996.
"trophic factors": Purves 1990.

86 *redundantly represented:* Here a cell assembly must be redefined as a set of neurons such that every connection between neurons is a strong synapse, provided that it exists. We could revise Locke's metaphor by imagining writing on white paper in which someone has randomly poked many holes (without trying to avoid the holes). The missing parts of the paper are analogous to the missing synapses in a sparsely connected network. If your handwriting is much bigger than the holes, the information may still be readable. But if your handwriting is too small, information will be lost.

87 *the method of loci:* Yates 1966.
"connections are created between them": It's this variant of Hebb's rule that is expressed by a ditty popular among college students learning neuroscience: "Neurons that fire together, wire together."

88 *Gerald Edelman:* Edelman 1987; Changeux 1985. A contrarian view was presented in Purves, White, and Riddle 1996, which was answered by Sporns et al. 1997.

89 *generated "on demand":* The "on demand" theory of synapse creation is analogous to Jean-Baptiste Lamarck's theory of evolution. Lamarck argued that animals can pass on acquired characteristics to their offspring, so that variation is adaptive rather than random. For example, he believed that a person who grows larger muscles through physical training can pass on larger muscles. Lamarck's ideas were discredited but have recently been partially revived by research on epigenetics.

Jeff Lichtman: Findings are reviewed in Lichtman and Colman 2000; a readable introduction to fundamental ideas is provided by Purves and Lichtman 1985.

90 *little or no memory loss:* Gilbert et al. 2000.

91 *below 18 degrees:* PHCA is sometimes used when neurosurgeons remove brain aneurysms. The circulation is stopped to prevent bleeding while the aneurysm is clipped, and the low temperature prevents the brain damage that would otherwise be caused by lack of oxygen during that time. At such low temperatures the heart doesn't beat properly; it is

stopped completely by injecting potassium chloride (one of the drugs used in execution by lethal injection).

two storage systems: Actually it's more complex than this, because there are additional information stores inside the microprocessor. The RAM and hard drive are just the off-board information stores.

92 *change more slowly:* I should mention that synapses also change their strengths more rapidly and temporarily. This is known as short-term plasticity, and could also be a basis of short-term memory.

6. The Forestry of the Genes

100 *different adoptive families:* Bouchard et al. 1990.

persons chosen at random: Strictly speaking, the proper comparison would be with two individuals drawn from different pairs of monozygotic twins raised apart.

101 *little room for argument:* In a celebrated case, Sir Cyril Burt, a pioneer in the study of twins, was posthumously accused of fabricating his data. This cast doubts on the whole field, which were eventually dispelled by more solid data.

First Law of Behavior Genetics: Turkheimer 2000. The Second Law is "The effect of being raised in the same family is smaller than the effect of genes," and the Third Law is "A substantial portion of the variation in complex human behavioral traits is not accounted for by the effects of genes or families."

If one twin has autism: Steffenburg et al. 1989; Bailey et al. 1995. There is a range because the exact numerical value depends on whether autism is defined strictly or, as in the autism spectrum disorders, more inclusively. Also, sample sizes are fairly small, so the numbers are subject to statistical uncertainty.

concordance rate for autism: Hallmayer et al. 2011 revises the concordance rate for DZ twins upward relative to the earlier studies of Steffenburg et al. 1989 and Bailey et al. 1995. According to the newer estimates, genetic influences are important for autism, but not as much as previously thought.

102 *What about schizophrenia?:* Cardno and Gottesman 2000.

synthesize proteins: You might have thought that cells don't have to make proteins because we ingest them from food. But actually the digestive system chops up proteins into amino acids, and our cells reassemble them into different proteins.

103 *contains the same genome:* There are some exceptions to this rule, such as certain cells in your immune system, variations arising from errors in DNA replication, and so-called mosaic organisms.

104 *inside and outside of the neuron:* Sometimes the door and tunnel are in a nearby molecule rather than in the receptor itself. The receptor can open the door by sending another signal, much as electrically powered doors are opened by pressing a button off to the side. Such a receptor is not an ion channel, and is said to be "metabotropic." The type of receptor discussed in the text is an ion channel, and is said to be "ionotropic."

called a "channelopathy": Kullmann 2010.

105 *clay or metal caps:* Miles and Beer 1996; Leroi 2006.

reduced brain size at birth: Brain size of at least two or three standard deviations below the norm is the clinical definition of microcephaly (Mochida and Walsh 2001).

pattern of folds: Mochida and Walsh 2001.

intermarriage between cousins: Leroi 2006; Mochida and Walsh 2001.

severe mental retardation: Mochida and Walsh 2004.

106 *control neuronal migration:* Guerrini and Parrini 2010.

growth cone acts like a dog: Kolodkin and Tessier-Lavigne 2011.

200 million axons: The numerical estimate comes from Tomasch 1954 and Aboitiz et al. 1992.

milder than in microcephaly: Paul et al. 2007. "Split-brain" patients, with a corpus callosum severed by epilepsy surgery, also have relatively minor impairments.

107 *half a million per second:* The estimate of half a million comes from Huttenlocher 1990, Figure 1, which summarizes data from Huttenlocher et al. 1982.

108 *number of synapses has dropped:* Huttenlocher and Dabholkar 1997. Similar observations were made in the monkey cortex by Rakic 1986.

110 *a less than ideal way:* Earlier I mentioned channelopathies, defective ion channels that cause electrical signaling of individual neurons and synapses to malfunction. Because neural activity alters connectomes by mechanisms like Hebbian plasticity, a channelopathy is expected to lead to abnormal connectivity. This example shows that connectopathies may be associated with other types of neuropathology.

111 *autistic brain is slightly smaller:* Redcay and Courchesne 2005 is a meta-analysis, combining the results of many studies.

schizophrenia, like autism: Lewis and Levitt 2002; Rapoport et al. 2005.

112 *too few connections:* Courchesne and Pierce 2005; Geschwind and Levitt 2007.

schizophrenia, too, be caused: Friston 1998. According to Kubicki et al. 2005, Carl Wernicke and the German psychiatrist Emil Kraepelin proposed the connectopathy theory of psychosis at the beginning of the twentieth century.

rapidly again in adolescence: Huttenlocher and Dabholkar 1997.

over the edge to psychosis: Is the connectopathy theory consistent with the observed effects of schizophrenia medications? Psychotic symptoms are relieved by drugs interfering with synapses that secrete dopamine. The symptoms are induced in normal people by drugs interfering with synapses that secrete glutamate. (Examples are ketamine and phencyclidine or PCP, which temporarily turn recreational users into schizophrenics, as emergency room physicians can attest.) According to the traditional view, the connectivity of the schizophrenic brain is normal, but the synapses don't work properly. Synaptic malfunction is corrected by the antipsychotic drugs and induced by the psychosis-generating drugs. But another view would be that antipsychotic drugs cause changes in synaptic function that compensate for connectopathy in schizophrenics, while psychosis-generating drugs mimic the effects of connectopathy in normals. This is possible because changes in synaptic function and changes in connectivity may have similar effects. For example, drastically weakening a synapse may be indistinguishable from removing it altogether. There is an even more subtle possibility. It may be wrong to think of abnormal synaptic function and abnormal connectivity as two independent defects. Suppose that synapse elimination is driven by synaptic weakening, which in turn is dependent on activity. If abnormal synaptic function changes activity patterns, it could end up causing the brain's connectivity to develop abnormally. Any initial abnormality in connectivity might also lead to abnormal activity patterns, which could cause further development of abnormal connectivity. Connectopathy would accompany schizophrenia, but it would be difficult to say which is cause and which is effect.

113 *predicting the IQ of individuals:* Actually neo-phrenologists can predict mental retardation with certainty in the special case that the brain is extremely small, as in microcephaly.

114 *will develop HD:* Since there is no cure, and since the test does not predict when the symptoms will start, most people with a family history of HD choose not to take the test.

genetics of autism and schizophrenia: Given enough time, genomics researchers may eventually identify all the different genetic defects involved in autism. Then perhaps a large battery of genetic tests will make it possible to predict autism accurately. But even if all the relevant mutations are known, the complex interactions between genes may still make it difficult to predict autism accurately.

correcting the genetic defect: Ehninger et al. 2011; Guy et al. 2011.

7. Renewing Our Potential

117 *5 or 10 percent:* The numbers depend on the exact definition of "long-term." In a more recent book, Seligman says that 5 to 20 percent of dieters regain their lost weight (or more) within three years (Seligman 2011).

"zero-to-three movement": Bruer 1999.

118 *elderly as well as young adults:* Draganski et al. 2004; Boyke et al. 2008.

videos of these remarkable processes: Meyer and Smith 2006; Ruthazer, Li, and Cline 2006.

119 *wires themselves are fixed:* It's common to use the term *rewiring* to include reconnection as well, but I think it's more helpful to distinguish between them.

brain into regions: Karl Lashley, the proponent of the principle of equipotentiality, was the most vigorous opponent of cortical localization. He had many ways of downplaying the existence of cortical areas. One of them was to deny or question whether localization had any significance for function: "The basis of localization of function within the nervous system is apparently the grouping of cells of similar function within brain regions. . . . What activities of the cells are favored by such an arrangement? What functions does it permit that could not be carried out if the cells were uniformly distributed throughout the system? Has localization or gross anatomic differentiation any functional significance whatever? . . . Increasing knowledge of the facts of cerebral localization has only emphasized ignorance of the real reason for any gross localization whatever." Lashley's questions are answered in this chapter.

between nearby neurons: Schüz et al. 2006.

120 *touch, temperature, and pain:* The relevant brain regions belong to an important structure called the thalamus. As a rule, the most direct pathways from all the sense organs to the neocortex pass through the thalamus, which is sometimes called the "gateway to the neocortex." The thalamus sits at the top of the brainstem, and is surrounded by the cerebrum. Some authorities include the thalamus as part of the brainstem, while others regard it as part of the diencephalon, also known as the interbrain.

Gerald Schneider: A major pathway for auditory information travels from the ears to the brainstem to the inferior colliculus to the medial geniculate nucleus (MGN) of the thalamus to the primary auditory cortex (Brodmann areas 41 and 42). A major pathway for visual information travels from the retina to the superior colliculus (SC). Schneider 1973 and Kalil and Schneider 1975 damaged the SC as well as the axons traveling from the inferior colliculus to the MGN. This diverted retinal axons from growing into the SC, rerouting them toward the MGN to fill the "vacuum" that had been created there. In effect, the researchers wired the eyes to the nominal auditory system.

visual cortex was disabled: Visual information travels not only to the SC but also along another pathway from the retina to the lateral geniculate nucleus (LGN) of the thalamus to the primary visual cortex (Brodmann area 17). The MGN and the LGN are analogous parts of the thalamus, serving hearing and vision respectively. Sur, Garraghty, and Roe 1988 disabled the visual cortex by damaging the LGN. Similar results were obtained in Schneider's hamsters by Frost et al. 2000.

when they read Braille: Sadato et al. 1996; Cohen et al. 1997.

121 *wiring between regions is selective:* This principle of "wiring economy" explains why most neural connections are between nearby neurons, and most areal connections between

nearby areas. The principle can be formalized as the postulate that the connectome is re-alized using the minimum length of wires (axons and dendrites). Theorists have used it to explain why nearby neurons tend to have similar functions, and why this rule is some-times violated by discontinuities in cortical maps. (See Chklovskii and Koulakov 2004.) Wiring economy is an important design principle for electrical engineers, too. One of their challenges is to arrange transistors on the surface of a silicon slab to minimize the length of wire required to establish a desired connectivity.

constrain the potential: This echoes the earlier discussion of memory, which argued that neurons are sparsely connected because full connectivity would be wasteful of space and other resources. I theorized that sparse connectivity constrains the potential of neurons to store new associations, and reconnection renews this potential.

feral children could not learn: The critical period applies only to the learning of a first lan-guage. A second language, although much easier to learn before puberty, is not impos-sible in adulthood.

122   *took a tragic turn:* Jones 1995.

real sentence structure: Rymer 1994.

123   *Antonella Antonini and Michael Stryker:* Antonini and Stryker 1993, 1996. They studied the axons entering V1 from the LGN, a brain region described in an earlier note.

124   *deprivation was ended early:* Their results don't entirely explain the critical period for vi-sual development. Binocular deprivation leads to an abnormal visual system, but LGN axons corresponding to both eyes remain normal, or even larger than normal. Perhaps some other kinds of connections are affected, but Antonini and Stryker were not able to see this.

Greenough and his colleagues: Greenough, Black, and Wallace 1987.

George Stratton: Stratton 1897a, 1897b.

125   *their pointing arm:* Bock and Kommerell 1986.

126   *This skewed behavior:* Knudsen and Knudsen 1990.

Kennard Principle: Schneider 1979.

exceptions are well-known: For example, if the brain damage is very early — just days after birth — the effects can be more severe later on (Kolb and Gibb 2007). A more conserva-tive reformulation is: The earlier the damage, the greater the reorganization of the brain. The reorganization may succeed in restoring function, or it may not.

127   *new branches can grow:* Yamahachi et al. 2009.

lifetime of stereo blindness: Susan Barry had surgery to correct her strabismus at age two. If that surgery had happened later, it's not clear her special stereo training would have been as effective in adulthood.

Researchers have employed: Vetencourt et al. 2008; He et al. 2006; Sale et al. 2007.

more optimistic message: Linkenhoker and Knudsen 2002.

128   *injury facilitates rewiring:* Carmichael 2006.

subtler kinds of rewiring: In the Knudsen experiments, rewiring could be seen relative to the map in the inferior colliculus. A similar strategy could work in sensory and motor ar-eas of the cortex, which generally contain analogous maps. Many other areas are not or-ganized according to such simple maps, however, so rewiring is more difficult to detect.

129   *No new neurons:* Rakic 1985 cemented the dogma.

Elizabeth Gould: Gould et al. 1999.

"most startling": Blakeslee 2000.

champion at self-repair: Taub 2004.

prevailed in the neocortex: Most of the evidence comes from monkeys, but Bhardwaj et al. 2006 additionally studied the human brain.

hippocampus and the olfactory bulb: Kornack and Rakic 1999, 2001. New neurons in these regions of the adult rat brain had previously been shown by Joseph Altman in the 1960s, but his pioneering discovery had been largely ignored by his colleagues.

"gateway" to memory: Kempermann 2002.

memories of smells: Lledo, Alonso, and Grubb 2006.

130 *fingers fused together:* Flatt 2005.

died as survived: Cowan et al. 1984.

wasteful to create: Buss, Sun, and Oppenheim 2006.

I'll call regeneration: When neuroscientists use the term *regeneration,* they are usually referring to the regrowth of axons after they are severed, but I call this *rewiring.* My usage of *regeneration* is typical of biology, and refers to the creation and elimination of cells.

since the 1960s: Gross 2000 reviews the history of such reports and speculates about why they were ignored.

grain of truth: Kornack and Rakic 2001 charged that Gould had erroneously identified non-neuronal cells as neurons. There are many types of brain cells that are not neurons.

131 *foster learning and plasticity:* On a related note, some critics say that the Rosenzweig experiments reveal the effects of deprivation, not enrichment. The fancy cages with toys and companions should not be regarded as "enriched," as they only relieve the deprivation of the ordinary laboratory cage. The latter is a highly impoverished environment compared with the rats' natural habitat.

132 *migrate into the zone:* Carmichael 2006.

8. Seeing Is Believing

138 *and Francis Crick:* Watson and Crick relied on the data of Rosalind Franklin, who was a crystallographer. She died prematurely and could not share their Nobel Prize.

141 *didn't fully recognize its significance:* Leeuwenhoek reported his observations of sperm in a letter to the president of the Royal Society of London. Embarrassed by the subject, he stressed that the specimen was the natural product of his marriage bed, and asked the president to suppress the letter if he found it offensive (Ruestow 1983).

called them "animalcules": Actually animalcules seem like an afterthought in the letter, because they are mentioned only in the last paragraph (Leeuwenhoek 1674).

three clergyman, a lawyer, and a physician: Dobell 1960 describes Leeuwenhoek's life and career, and collects many of his letters.

142 *single, very powerful lens:* Ford 1985 describes the history of the single-lens microscope and argues that Leeuwenhoek made his best lenses by letting molten glass solidify into small globules. Ruestow 1996 notes that Leeuwenhoek also made some lenses by the more standard method of grinding glass, as he claimed in his writings.

individual neurons: Figure 26 depicts Golgi-stained neurons from the cortex (superior temporal sulcus) of the adult rhesus monkey. The image extends from the white matter at the bottom to layer 3 of the cortex at the top, a distance of roughly 1.5 millimeters.

143 *single dark strand:* Those of you who are observant may notice that the pasta shown in Figure 27 is actually bucatini, which is thicker than spaghetti and has a hole running down the center. (It has a wonderful chewy texture, and I recommend it highly.) If every strand of the bucatini were stained with a unique color, it might be possible to trace the paths of *all* the strands, even in a somewhat blurry image. Researchers have actually implemented this strategy by genetically engineering mouse neurons to fluoresce in random colors, a method that Jeff Lichtman wittily named "Brainbow" (Livet 2007; Lichtman 2008). However, the number of distinguishable colors is limited, so Brainbow may be insufficient for tracing a large number of densely entangled neurites. It may be possible to improve the situation by combining Brainbow with sharper images, like those produced by recently invented methods of light microscopy that beat the diffraction limit (Hell 2007). In an-

other approach, Tony Zador has proposed genetically engineering each neuron to contain a random RNA or DNA sequence. The sequence could be unique for every neuron, because the number of possibilities is so large — much larger than the number of distinguishable colors. Other molecular tricks and genomic technologies would be used to find the sequences for every pair of connected neurons, yielding the connectome. We don't yet know whether these directions of research will provide alternatives to electron microscopy, the standard method of finding connectomes. I mention them only to make clear that connectomics is going through an exciting period of innovation.

why Golgi's stain: From a solution of potassium dichromate and silver nitrate, silver chromate precipitates in a small fraction of neurons, for some unknown reason.

144 *Golgi looked in* his *microscope:* Guillery 2005. Cajal's view was called the "neuron doctrine" and Golgi's the "reticular theory."

145 *as Golgi envisioned:* Have you ever heard the joke "Economics is the only field in which two people can share a Nobel Prize for saying opposing things"? The quip probably dates from 1974, when the prize was shared by the economists Gunnar Myrdal and Friedrich Hayek, who were shocked to find themselves honored at the same event, given that their views were so diametrically opposed. At his banquet speech Hayek suggested that a prize for economics was a bit dangerous. Myrdal even wrote a paper calling for the abolition of the prize (Myrdal 1977). He argued that economics was a "soft" science, so its prize, established in 1968, did not belong with the "real" Nobel prizes in the "hard" sciences, which were originally established by the will of Alfred Nobel in 1895. According to Lindbeck 1985, this was ironic coming from Myrdal, who had lobbied strongly for the creation of the economics prize in the first place. Based on the 1906 Nobel Prize to Golgi and Cajal, should we also regard neuroscience as a "soft" science? Perhaps neuroscience is somewhere in between economics and physics. It's true that Golgi and Cajal had opposing views, but no one called for the abolition of the Nobel Prize for Physiology and Medicine, as far as I know. And they both turned out to be correct, so the Nobel committee did the right thing.

new stains: These are based on big and heavy atoms like osmium, uranium, and lead, which reflect electrons well.

Figure 28: This transmission electron microscope image comes from the rat hippocampus. It can be found along with many other interesting images of neurons and synapses at synapse-web.org.

146 *the diffraction limit:* Recently physicists have realized that it's possible to beat the diffraction limit using fluorescence microscopy, which was not available to Golgi (Hell 2007).

in a light microscope: The blurred version of the image is due to Winfried Denk, who simulated the point-spread function of a 1.4 numerical aperture (NA) microscope objective assuming a wavelength of 500 nanometers.

147 *edge of a saw is blunt:* As a hybrid of saw and knife, serrated knives are one of those irritating intermediate cases that are the bane of the classifier. We will ignore them.

2 nanometers wide: More precisely, 2 nanometers is the edge radius of curvature claimed by several manufacturers of diamond knives on their websites. In the published literature, one can find reports of 4 nanometers (Matzelle et al. 2003).

148 *Keith Porter and Joseph Blum:* Porter and Blum 1953. Bechtel 2006 recounts the history of biological electron microscopy.

150 *ultramicrotome mounted inside:* Denk and Horstmann 2004.

"scanning electron microscopy": Earlier researchers had used transmission electron microscopy (TEM), which sends electrons through thin slices of tissue. (This is similar to viewing a photographic negative by holding it up to a light.) The scanning electron microscope instead bounces electrons off the surface of the object being imaged.

151 *thin as 25 nanometers:* This number is important, because it sets the resolution of the

3D image stack in the vertical direction. Electron microscopy has much finer resolution (nanometers or less) in the two lateral directions. The vertical resolution is much coarser.

154 *eventually achieved 30 nanometers:* Hayworth's original design, shown in Figure 30, was called ATLUM rather than ATUM. The *L* stood for "lathe," a kind of rotary machine tool. The plastic block containing brain tissue was mounted on an axle. Each turn of the axle pushed the block past the diamond knife, shaving a thin slice off. Hayworth initially thought that the rotary motion would control slice thickness more precisely. Since then, he has returned to the traditional linear motion of a conventional ultramicrotome, like the back-and-forth of meat in a deli slicer.

eliminates the need for a diamond knife: Knott et al. 2008 describes the method of focused ion beam (FIB) milling. Bock et al. 2011 describes a modification of the transmission electron microscope to produce images with larger field of view, speeding up the rate of data acquisition.

9. Following the Trail

156 *walls of the axon go by:* The molecular car is kinesin, and the track is called a microtubule.
one billion collisions: CMS Collaboration 2008.

157 *had not been invented yet:* When Brenner gave the kickoff lecture for my 2007 class on connectomics, he expressed disdain for the term. He recommended that the field be christened "neuronomy" instead, quipping that "neuronomy is to neurology as astronomy is to astrology."

158 *Richard Goldschmidt:* White et al. 1986.

159 *sausage is stuffed with spaghetti:* We are stretching our Italian food analogy. Perhaps Thai food would be better, as summer rolls typically do contain noodles.
just thin enough: Ideally, the slice thickness would be the same as the spatial resolution of the 2D images produced by the electron microscope. Then the 3D image would have the same spatial resolution in all directions. But it's not possible to slice that thin, so the image inevitably has poorer resolution in the third dimension.

160 *repeatedly wrote the same symbol:* They wrote with felt-tip pens on transparent acetate sheets, which were placed on top of the original photographic plates. To make the process even more complex, sometimes they would trace two neurites that started out separate but merged at a branch point. Once they realized the two neurites were part of the same neuron, they went back and changed all the letters of one neurite to match the other.
302 neurons of the worm: To be more precise, the 282 somatic neurons are described. There are also 20 pharyngeal neurons, which form an almost independent nervous system (Albertson and Thomson 1976). Errors were corrected, inconsistencies resolved, and gaps filled in by Chen, Hall, and Chklovskii 2006. The updated version was published at wormatlas.org.

161 *touch to the head:* Chalfie et al. 1985.
John Fiala and Kristen Harris: Fiala 2005.
render parts: The thicker object is a short segment of a dendrite, with spines protruding. The thinner objects are parts of axons.

162 *a million person-years:* Helmstaedter, Briggman, and Denk 2008.

165 *machine learning:* How can you help a computer learn? First, devise an algorithm that performs the task, but put a lot of adjustable parameters in it. Depending on the parameter settings, the algorithm performs the task differently. Second, devise a quantitative measure of disagreement between the computer and the humans on the database of examples. This measure is a function of the adjustable parameters in the computer program. It is known as a cost function, or objective function for learning. We would like to minimize this function with respect to the adjustable parameters. To do this, we carry out

the third and final step of writing a program that searches for the optimal setting of the parameters. Often this is done in an iterative fashion. The program finds a small change of the parameters that lowers the cost function. It does this repeatedly, in an attempt to find the lowest possible value.

Viren Jain and Srini Turaga: Jain, Seung, and Turaga 2010.

167   *Old-time computer hackers:* Kelly 1994.

     Intelligence Amplification: Engelbart credited the term to Ross Ashby, a pioneer in the field of cybernetics.

     possible to "crowdsource": I haven't mentioned that humans also make errors when tracing neurites, though at a lower rate than computers. Helmstaedter, Briggman, and Denk 2011 shows how to combine the efforts of multiple humans to improve accuracy, an example of the "wisdom of crowds."

169   *cost per letter:* Shendure et al. 2004.

10. Carving

171   *brain forest:* Cajal may have originated the metaphor, describing the brain as "a jungle, in whose impenetrable thickets many explorers had lost their way" (Ramón y Cajal 1989).

172   *Huntington's disease:* Utter and Basso 2008.

     it is very important: In this book I've been guilty of cortical chauvinism. For the sake of simplicity, I've spoken of localizing mental functions within cortical areas, but this is admittedly naïve. Every other brain region has its partisans, who can explain why the region is so important, even if smaller than the cortex. Fans of the basal ganglia have mapped its connections with the cortex and the thalamus to understand how these regions cooperate to carry out mental functions (Middleton and Strick 2000).

173   *Each strip:* Masland 2001. The figure presents a classification of neurons valid for a generic mammalian retina. Some larger types of neurons are omitted. I've used the terms *class* and *type* to denote two levels of taxonomy, but my usage is by no means standard in neuroscience. To classify plants and animals, biologists use the official terms *species, genus, family, order,* and so on. A similar scheme is needed for neurons.

174   *make out six layers:* According to the standard convention, most of the cortex has six layers, and is called *neocortex* or *isocortex.* "Neo-" refers to the evolutionary theory that six-layered cortex is newest. Those who don't believe this theory prefer the prefix "iso-", which emphasizes that all six-layered cortex has a similar appearance. Other parts of the cortex have fewer (or more) than six layers and are known as *allocortex.* A famous example is the hippocampus.

     layering was uniform: The arrangement of cell bodies into layers is known as *cytoarchitecture,* as "cyto-" means "cell."

175   *Oskar and Cécile Vogt:* Zilles and Amunts 2010. Their stain marked a substance called myelin, a fatty material that sheaths many axons. This revealed "myeloarchitecture" rather than the "cytoarchitecture" used by Brodmann.

     Sir Grafton Smith: Smith was an interesting character who straddled the fields of neuroanatomy and archaeology. He investigated and x-rayed the brains of Egyptian mummies.

     Percival Bailey and Gerhardt von Bonin: Bailey and von Bonin used a "double-blind" method to see whether cortical areas could be reliably distinguished by cytoarchitecture, and mostly found negative results (Bailey and von Bonin 1951).

176   *hundreds of neuron types:* Stevens 1998.

     Neuroscientists continue to argue: Nelson, Sugino, and Hempel 2006.

177   *A great many wires:* I should qualify my statement. It might make sense to regard the grooves (sulci) in the cortex as its "joints." Neurons on opposite sides of a groove are connected by longer axons than neurons within the same convolution (gyrus). By the princi-

ple of wiring economy, there should be fewer wires connecting opposite sides of a groove, and cutting along the grooves is analogous to carving at joints. This justifies the practice of MRI researchers, who locate cortical areas relative to grooves, because they can't see the layering that Brodmann relied upon.

Unlike poultry: If we want to preserve Socrates' metaphor, we can think of classification as happening by cuts in a high-dimensional feature space rather than three-dimensional space.

over one hundred types: White et al. 1986.

Collapse all neurons of one type: Ibid.

179 *Connections are directly related:* As described by Nelson, Sugino, and Hempel 2006, it's also important to define neuron types by molecular criteria such as the expression of a particular gene or genes. A beautiful example in the retina is provided by Kim et al. 2008. The molecular definition is useful for controlling neuron types and for understanding how they emerge during development. As I mentioned earlier, neurons of one type should also share similar functions, as revealed by measurements of spiking. Therefore, I anticipate three definitions of neuron types based on molecules, connectivity, and activity, which will ideally coincide with each other. These three definitions parallel three meanings of the term *neuron,* which were delineated by Golgi in his Nobel lecture. He pointed out that the neuron is supposed to be an embryological, anatomical, and functional unit, before he proceeded to question its existence.

Layer 4 of area 17: The axons reaching layer 4 of area 17 come from neurons in the LGN, which in turn receives axons from the retina. The LGN is a subdivision of the thalamus devoted to vision. As a general rule, sensory pathways reach the neocortex through thalamic axons terminating in layer 4. The text focuses on connections between areas, but differences in layering also reflect connections between neurons in the same cortical area, because of rules of connection that are based on layers. For example, excitatory neurons in layer 4 make synapses on pyramidal neurons in layers 2 and 3, which in turn make synapses on pyramidal neurons in layer 5. Therefore, when the thickness and density of layers change, connectivity is probably changing too.

area 17 has a thicker layer 4: Furthermore, there is potentially much more information in connectivity than in layering. Brodmann and his contemporaries disagreed over their cortical maps precisely because differences in layering are so subtle. Cortical layers are not very distinct in the first place, as we saw earlier, and variations in them are even less distinct. I predict that differences in connectivity will be much more marked.

regional or neuron type connectome: You may find it confusing that by now I've defined three kinds of connectome. According to Lederberg and McCray 2001, the term *genome* also has multiple meanings. When first coined in 1920, it referred to the totality of chromosomes in an organism. (Your DNA is divided into twenty-three pairs of molecules known as chromosomes, which are like volumes of an encyclopedia.) Later it came to refer to the totality of genes, and today it means all the letters in the DNA sequence. Similarly, I expect that the most common meaning of *connectome* will shift over time toward the neuronal one, which has the highest resolution.

180 *Wernicke called it:* Eling 1994.

181 *different flavor from the neural:* Catani and ffytche 2005; Mesulam 1998; Geschwind, 1965a, 1965b.

Olaf Sporns and his colleagues: Sporns, Tononi, and Kotter 2005. Around the same time, Patric Hagmann independently coined the term in his Ph.D. thesis.

182 *lesions that spare:* Mohr 1976.

less localized than previously: Lieberman 2002; Poeppel and Hickok 2004; Rilling 2008.

deny that the arcuate fasciculus: Bernal and Altman 2010.

other pathways that do: Friederici 2009.

the Broca–Wernicke model: Hickok and Poeppel 2007.

184 *formation of cortical areas:* Fukuchi-Shimogori and Grove 2001.

11. Codebreaking

185 *Michael Ventris and John Chadwick:* Chadwick 1960 recounts the story of their collaboration. Kahn 1967 tells a shorter version, along with providing other examples of codebreaking throughout history.

a number of lost languages: Robinson 2002.

186 *stick in your computer:* A similar scenario is the basis of Anthony Doerr's short story "Memory Wall."

187 *name the president:* Corkin 2002.

MTL seemed essential: In technical terms, H.M.'s condition is described as severe anterograde amnesia. "Anterograde" means that his amnesia only applied to events after his operation. His memory of events from before his surgery was mainly intact, though it was worse for events just prior to the surgery than from events long before. Therefore he had a mild degree of retrograde amnesia, which was temporally graded.

role in memory recall: Gelbard-Sagiv et al. 2008.

188 *groups of CA3 neurons:* This idea is due to David Marr, who first theorized about cell assemblies in CA3. Neurons in other parts of the hippocampus make synapses on neurons in other brain regions, rather than their neighbors.

the human CA3: Furthermore, it's not clear whether memories and cell assemblies are really confined to CA3. This might be the case for new memories, if they are initially stored in the hippocampus and later transferred to the neocortex, as some theorists believe. Alternatively, cell assemblies might be distributed across both the hippocampus and the neocortex from the very beginning. They might start out with more neurons in the hippocampus but end up with more neurons in the neocortex as memories are consolidated.

It includes autobiographical: These are called "episodic" and "semantic" types of memory, respectively. H.M.'s semantic memory was not as impaired as his episodic memory.

189 *declarative memory in animals:* Declarative memory might seem to depend on language, as the term implies that recall occurs through "declaring." But Eichenbaum 2000 argues that the term should nevertheless be extended to animals, because they have mnemonic capabilities that correspond to those included in human declarative memory and depend on analogous brain regions. Also, parrots and other animals might be able to "declare" memories through vocalization or other communication skills.

don't nurse their young: You might think that egg-laying distinguishes birds from mammals, but a few mammalian species like the platypus also lay eggs.

Males of other species also sing: Not all sounds from birds are considered song. Less complex sounds are known as "calls."

Mozart kept a pet starling: West and King 1990.

190 *starts to "babble":* Doupe and Kuhl 1999.

learns to copy his father's: If young zebra finches do not hear an adult male's song, they will still sing when they grow up, but abnormally. However, Fehér et al. 2009 showed that if such isolated birds are bred for several generations, so that each generation learns from the previous one, the song eventually sounds more normal again. This suggests that there is some innate preference for certain song properties, in addition to the preference learned from experience.

muscles around the syrinx: Also involved are the muscles for respiration, which control the rate of air flow through the syrinx.

191 *converted into sounds:* It's admittedly simplistic to say that RA and nXII merely relay or

amplify signals. For a more accurate account, you can consult the scientific literature. Also, you might question whether a straight pathway is a good model. Since the bird hears its own song, maybe there should be an additional step from the syrinx back into the brain, which would turn the pathway into a circular loop. In this view, each note of the song would serve as a stimulus that drives the bird to produce the next note. Such a loop was proposed as a model for sequence generation in the nineteenth century, by people like the American psychologist William James. It does not appear to be a good model for birdsong, because adult zebra finches can still sing even if they are deaf.

letters stand for nothing: Jarvis et al. 2005.

192 *dorsal ventricular ridge:* Karten 1997.

Michale Fee and his collaborators: Hahnloser, Kozhevnikov, and Fee 2002.

expect from a synaptic chain: The synaptic chain is actually a bit too simple a model for HVC. To account for the repetitions of the song motif, the last neurons in the chain would have to make synapses onto the first neurons, creating a circular structure rather than a linear one. And some additional mechanism would be needed to terminate the sequence after a few repetitions.

193 *like a synaptic chain:* Fee and his collaborators estimate that one hundred RA-projecting HVC neurons are spiking during any moment of song (Fee, Kozhevnikov, and Hahnloser 2004) and hypothesize that HVC contains a synaptic chain with one hundred neurons in each link.

To reveal it: Ideally, the HVC connectome would come to us naturally unscrambled, so no additional work would be necessary. This would be the case if HVC neurons were arranged so that they spiked in some spatially defined order — for example, from front to back. But actually it appears that neurons are arranged without regard to their spike times (Fee, Kozhevnikov, and Hahnloser 2004).

computer would be necessary: Actually, we could still do it by hand if the chain were perfect. But if there are some "inappropriate" connections, such as synapses directed backward, finding a chain becomes more difficult and requires a computer (Seung 2009). Unscrambling neurons is an example of a problem called "graph layout" by computer scientists.

194 *resemble blinking lights:* These stains fluoresce when illuminated, like a sticker that glows in the dark when illuminated by black light. The amount of fluorescence varies with calcium concentration, which in turn is modulated by spiking.

195 *might not be able to order:* Actually, this outcome could leave ambiguity. Perhaps a sequential ordering exists, but our unscrambling algorithms are too poor to find it. Computer scientists will have to work hard to make sure that their algorithms are good enough to find any ordering if it exists.

go backward or jump: Even if there turn out to be some "inappropriate" connections that violate the sequential ordering, we could still say that the connectome is an *approximation* to a synaptic chain. But if there were too many such connections, then we'd have to say that the chain is a bad model and cannot explain why the network generates sequential activity.

HVC neurons in young males: Jun and Jin 2007; Fiete et al. 2010.

reconnection also plays a role: This was suggested by Jun and Jin 2007.

198 *Kevin Briggman:* Briggman, Helmstaedter, and Denk 2011.

Davi Bock: Bock et al. 2011.

great-great-grandma's dog: What about grounding the memory of the bird's song? If we found an entire bird connectome, we could examine the pathways from each HVC neuron to the vocal muscles. These pathways are thought to transform the abstract sequence in HVC into the specific motor commands required to make sounds. (This transformation appears to be learned by practice too.) Analysis of the connections in these pathways might make it possible to decode the movement signaled by each HVC neuron. This method would require that we identify rules of connection for neurons related to motor control, which are analogous to the part–whole rule for perceptual neurons. In general, grounding memories requires that we trace pathways all the way from the center of the brain to the sensory and motor periphery.

199　*rules of connection:* Rules of connection can be mathematically formalized as probabilistic models of graph generation based on latent variables at the nodes of the graph (Seung 2009).

　　quite improbable too: Mooney and Prather 2005.

12. Comparing

201　*Native American and African myths:* Davis 2005.

　　bedrock assumption: Even more disconcertingly, identical twins challenge the more sweeping axiom that everything is unique — human, animal, or inanimate object. This axiom underlies the lovely claim that no two snowflakes are alike, and may have been behind the animistic beliefs of primitive societies that all objects have souls. Because of mass production by factories, we have grown blasé about material objects that look almost indistinguishable. Such instances were much rarer in the preindustrial world, so I suspect that twins appeared even more magical to our primitive ancestors than they do now. But such thoughts are less relevant for connectomics than fodder for nanotechnologists who promise to make material objects that are truly identical, down to the placement of individual atoms (see, for example, Drexler 1986).

202　*deviations in DNA sequence:* Machin 2009 discusses both genetic and epigenetic differences between identical twins.

203　*two complete* C. elegans *connectomes:* As mentioned earlier, the researchers actually pieced together the connectome using images drawn from several worms. The published *C. elegans* connectome is a mosaic, not a unified representation of an individual worm's nervous system. So we don't have even one complete connectome of an individual worm, much less two.

　　David Hall and Richard Russell: Hall and Russell 1991.

　　purebred dogs and horses: Laboratory animals are generally inbred this way to ensure that they are genetically almost identical, which is supposed to make experiments more repeatable. It's well-known that inbreeding can increase the likelihood of having two defective copies of a gene, and "recessive" disorders are governed by a "two strikes and you're out" rule. This is why many dog breeds have genetic disorders and why European royalty suffered from hemophilia. Since inbreeding probably makes laboratory animals "dumber," research on them might not be applicable to their wild counterparts.

204　*sophisticated computational methods:* The most basic computational problem of genomics is finding a matching or alignment between two DNA sequences. This is solved by fast approximations to dynamic programming, a formalism first developed in the 1940s and 1950s for solving problems with a one-dimensional or tree structure. Solving the analogous matching problem for two connectomes will be an important computational challenge for connectomics, and is much harder than aligning genomes. Determining whether two connectomes are the same is known as the graph isomorphism problem, for which no polynomial time algorithm is known. Determining whether one connectome is part of another is known as the subgraph isomorphism problem, which is NP-complete.

205　*known in antiquity:* Gray and white are not the natural colors of living brain tissue, which is pinkish, but rather the colors of preserved brain tissue.

is all "wires": As noted by Kostovic and Rakic 1980, Cajal already observed that there are exceptions to this rule, known as "interstitial neurons."

206 *straight out of the base:* This mental picture is a bit confusing, because the cell body looks like an arrowhead pointing in the opposite direction of information flow along the axon.

207 *150,000 kilometers:* This crude estimate assumes that the density of axons throughout the cerebral white matter is the same as in the corpus callosum, or 380,000 axons per square millimeter (Aboitiz et al. 1992). The estimate also makes use of the total volume of white matter, which is 400 cubic centimeters (Rilling and Insel 1999).

208 *Myelination speeds up:* The fat in myelin serves as an insulator that prevents leakage of electrical currents out of the axon. This has the effect of boosting the speed at which electrical signals propagate. Electrical signals travel at top speed in myelinated axons, ten or more times faster than in unmyelinated axons. Myelin sheaths are outgrowths of nonneuronal, or glial, cells. Schwann cells myelinate PNS axons, and oligodendrocytes myelinate CNS axons.

axon enters and branches: If the axon doesn't branch in a region, it's probably passing through without making synapses.

209 *almost completely unexplored:* Historically, the white matter of animal brains has been studied by the method of tracer injection. When certain substances are injected into the brain, they are taken up by neurons at that location and transported along axons to other brain regions. By visualizing the destination of such tracer substances, it is possible to identify the regions connected to the injection site. Data from such experiments was compiled in Felleman and Van Essen 1991 to chart the regional connectome of the monkey brain shown here (Figure 51). The Brain Architecture Project, led by Partha Mitra, is systematically applying tracer injections with the goal of producing a complete map of long-range connections in the rodent brain. But the tracer must be injected while the brain is still alive, as its transport depends on active processes in living neurons. Therefore tracer injection is an invasive technique, and is employed only with animal brains. It does not work at all with postmortem human brains. (Certain lipophilic dyes don't depend on active transport, but are difficult to use as tracers in postmortem brains because they travel so slowly.) My proposal of serial light microscopy does not require injection of tracers. Instead of staining just a small bundle of axons, all myelinated axons in the white matter are stained and imaged. This method could potentially be applied to a postmortem human brain. Furthermore, its high spatial resolution prevents the ambiguities that plague diffusion MRI and naked-eye dissection. My proposal is an example of dense reconstruction, which extracts a complete map from a single brain, rather than aggregating data from many brains.

210 *Diffusion MRI is an exciting:* This method works by measuring the direction dependence of the speed of diffusion of water molecules in the brain. Diffusion along the axis of axons is faster than in the perpendicular direction.

211 *sparking revisions:* Friederici 2009.

212 *complementary methods:* We've focused on comparing connectomes of different individuals using microscopy. This provides snapshots of connectomes at moments in time. Comparing such snapshots can tell us something about how interventions change the brain. (Recall that Rosenzweig's experiments on environmental enrichment and Antonini and Stryker's experiments on monocular deprivation of V1 relied on comparisons between different animals or populations of animals.) But we would also like to compare the connectomes of a single individual at different times. Unfortunately, there is currently no good way of doing this. A noninvasive method like MRI can follow the evolution of a connectome over time but cannot deliver the neuronal resolution of microscopy. There are ways of improving the snapshots of microscopy by highlighting changes to the con-

nectome, however. There now exist staining methods for making recently strengthened synapses visible, as well as methods that do the same for newly created neurons. It's important to invent ways of labeling *synapses* that were recently created, as well as locations where synapses were recently eliminated. With such images, one could not only quantify the total amount of synapse creation and elimination but go much further, because every created and eliminated synapse would be seen in the context of an entire network. We would know exactly how synapse creation and elimination changed the organization of connectivity, as opposed to a coarse measure like total number of synapses. This would enable us to detect even subtle connectome changes, as well as figure out whether they are causally related to learning.

brains of the deceased: I mentioned earlier that the two-photon microscope can be used to observe neurons in living brains. This requires opening or thinning the skull, however. Also, it works only for neurons near the surface of the brain, unless the viewing is done through an optical fiber inserted deep inside, an even more invasive procedure. And it can visualize only neurites that are sparsely labeled.

present special problems: The brains may not be well preserved after death; they may suffer from other abnormalities that are not relevant to the mental disorder in question, such as injury caused by stroke; and they may have been changed by drugs if the deceased person was treated for the mental disorder.

into the genomes of animals: Nestler and Hyman 2010. Some mental disorders are associated with deletions of parts of the genome, and researchers can create these deletions in animal genomes also.

simian immunodeficiency virus: According to one theory, HIV originated when SIV mutated and jumped from monkeys to humans.

213 *numbers of plaques and tangles:* Oddo et al. 2003.

214 *"unbiased, hypothesis-free manner":* Lander 2011.

stroke of insight: Other times, it's the available tools of measurement that motivate the hypothesis. For example, Galton hypothesized that intelligence was related to head size mainly because he was able to measure head size, not because this was a great hypothesis.

13. Changing

216 *Der Freischütz:* This literally means "The Freeshooter," but it's typically translated as "The Marksman."

suffering of millions of people: Bosch and Rosich 2008.

inspired by Weber's popular opera: Strebhardt and Ullrich 2008. Ehrlich also invented the idea of receptor molecules.

217 *last-ditch measure:* The current practice of psychosurgery and the history of the "frontal lobotomy," which earned the Portuguese physician Egas Moniz a Nobel Prize in 1949, are described in Mashour, Walker, and Martuza 2005. While lobotomy could reduce the symptoms of psychosis, it also mentally crippled the patients. It became apparent that the side effects were worse than the disease. Because of psychosurgery's abuses, many regard the prize to Moniz as an embarrassment to the Nobel committee. However, some historians argue that psychosurgery was justifiable in an age before antipsychotic drugs, when the only alternative was confinement in a mental institution. Much of the infamy of the procedure was due to American physician Walter Freeman, who developed a version of the procedure that he called the "transorbital leucotomy." In his gruesome technique—nicknamed the "ice pick lobotomy"—a mallet was used to drive a sharp instrument resembling an ice pick past the eye through the eye socket into the brain. Moving the tip back and forth destroyed tissue in the frontal lobe. Freeman's innovation made the

procedure so quick and easy that non-surgeons and even non-physicians could perform it.

218 *surplus or deficiency of neurotransmitter:* Schildkraut 1965.

effects of fluoxetine on the four R's: Hajszan, MacLusky, and Leranth 2005 found an increase in dendritic spine density, a sign of synapse creation. Wang et al. 2008 demonstrated increased dendritic growth of newborn neurons. The extensive literature on neuron creation in the hippocampus and its role in depression is reviewed in Sahay and Hen 2008.

specifically target connectomes: Other treatments for brain disorders involve manipulating neural activity. In electroconvulsive therapy (ECT), shocks administered through scalp electrodes induce epileptic seizures. ECT is far from a magic bullet, as the seizures spread unselectively over the brain, yet for some unknown reason ECT can alleviate symptoms of depression and other mental disorders. Better-targeted electrical stimulation can be performed using electrodes that are surgically implanted inside the brain. Symptoms of Parkinson's disease, for example, can be relieved by stimulating parts of the basal ganglia. Some researchers are developing even more precise therapies based on optogenetics, the optical stimulation of activity in a single neuron type that has been genetically altered to be sensitive to light. Like altering neurotransmitter levels, manipulating neural activity may sound completely different from promoting connectome change, but it's not. For example, the seizures of ECT may change the connectome through Hebbian plasticity, and it's quite possible that such changes are responsible for its therapeutic effects (and for side effects like amnesia).

supplemented by training regimens: It's intuitively plausible that combining drugs and "talk therapy" might be more effective than either alone. Evidence supporting this idea for the treatment of depression is presented by Keller et al. 2000.

219 *alive but damaged:* Lipton 1999.

"gene therapy" for Parkinson's: Yamada, Mizuno, and Mochizuki 2005; Mochizuki 2009.

degeneration in neurons: Some researchers report that neurons "die backwards" in many diseases. In other words, degeneration affects the synapses and tips of axons first, and then moves backward along the axon toward the cell body. The collapse of the axon in turn might trigger the neuron to initiate the suicidal mechanisms of programmed cell death. See Coleman 2005; Conforti, Adalbert, and Coleman et al. 2007.

220 *connections are lost:* Selkoe 2002.

before the first onset: Baum and Walker 1995.

221 *such as lizards:* Lledo, Alonso, and Grubb 2006.

fingertips grow back: Illingworth 1974.

Injury naturally activates: Carmichael 2006.

divert them from: Zhang, Zhang, and Chopp 2005.

survive in recipients' brains: Mendez et al. 2008.

222 *whether the transplants actually:* Olanow et al. 2003.

"reprogrammed" to divide: This is known as a patient-derived induced pluripotent stem cell (iPSC).

skin cells of Parkinson's: Soldner et al. 2009.

Whether created naturally: Zhang, Zhang, and Chopp 2005; Buss 2006; Lledo 2006.

added by transplantation: Brundin 2000.

molecules that promote plasticity: Murphy and Corbett 2009.

grow new axonal branches: Carmichael 2006.

223 *natural molecular processes:* Carmichael 2006. Reweighting might also be important for recovery from stroke, by unmasking previously nonfunctional pathways through strengthening of their synapses. Another type of change can unmask pathways, which should perhaps be included in reweighting. This is a change in the threshold for produc-

ing a spike. (In the weighted voting model, the threshold specifies the margin between "yes" and "no" votes required from presynaptic "advisors" for a neuron to fire an action potential.) Lowering thresholds can unmask pathways by making neurons more excitable, that is, less choosy about when to spike. This could be especially important for recovery from stroke, because the death of neurons reduces the number of advisors for the surviving neurons. They may receive less "yes" votes, so they will not spike unless their thresholds are lowered.

effect on learning and memory: Nehlig 2010.

deprived of cigarettes: Newhouse, Potter, and Singh 2004.

224   *nine out of ten:* Kola and Landis 2004.

a billion dollars: Morgan et al. 2011. These estimates are uncertain because such financial information is proprietary. Also, pharmaceutical companies have an interest in overstating their costs, to answer criticisms that they are greedily overcharging for their products.

swept through the psychiatric hospitals: The serendipitous history of antipsychotic drugs is reviewed in Shen 1999. The first-generation, or "typical," drugs were created by varying the molecular structure of chlorpromazine. The second-generation, or "atypical," drugs have more diverse molecular structures.

first antidepressant medications: Lopez-Munoz and Alamo 2009. Iproniazid was the first of the monoamine oxidase inhibitors, and imipramine kicked off the discovery of many tricyclic antidepressants.

225   *golden age of the 1950s:* Since the 1950s, the only major success story has been fluoxetine, which was discovered by rational means rather than serendipity. From studies of the first antidepressants, scientists had formulated the theory that depression had something to do with the brain system that secretes the neurotransmitter serotonin. In the early 1970s, the company Eli Lilly searched for molecules that acted on the serotonin system but lacked the side effects of the tricyclic antidepressants like imipramine. The search turned up fluoxetine, which was finally approved by the U.S. government in 1987. See Lopez-Munoz and Alamo 2009.

A drug is an artificial molecule: The line between artificial and natural is blurred by "biologics." Vaccines are the classic example, but newer ones are proteins identical or similar to the ones that occur naturally in the body. These can still be viewed as artificial, in the sense that they are synthesized or introduced by non-natural means. Biologics are distinguished from "small molecules," which contain many fewer atoms and are the classic kind of drug.

226   *the attrition rate:* Kola and Landis 2004.

between the first and last stages: Markou et al. 2008.

227   *humanized mouse models:* Legrand et al. 2009.

animal behavior that is analogous: Nestler and Hyman 2010.

14. To Freeze or to Pickle?

233   *probability theory:* The founding of probability theory is recorded in a series of letters between Pascal and another famous mathematician, Pierre de Fermat. See Devlin 2010.

a searing religious vision: The two-hour vision took place the evening of November 23, 1654, which has come to be known as Pascal's "night of fire." We know of it only because Pascal recorded the event on a document that was sewn into his coat, and discovered by a housekeeper after his death. See O'Connell 1997.

235   *one thousand living members:* According to an Alcor web page, as of July 31, 2011, there are 955 members and 106 cryopreserved "patients."

237   *want to live forever:* Some friends tell me that they wouldn't want to be immortal. This position has also been argued by philosophers, notably Charles Hartshorne. I find this

ironic, as I saw Hartshorne a few times in my father's office at the University of Texas, and he seemed practically immortal — he rode a bicycle well into his eighties and lived until age 103. But I agree with Camus that suicide is more interesting as a philosophical problem, since immortality doesn't seem like a realistic option anyway.

apocryphal, alas: Peck 1998.

the court sorcerer Xu Fu: Howland 1996.

238 *laid the athlete's remains to rest:* The Ted Williams story is told in Johnson and Baldyga 2009.

"Miracle of the Sun": There are many books on the miracle. Bertone and De Carli 2008 was written by a cardinal and endorsed by the pope. The apparition of the Virgin Mary revealed three secrets to the shepherd children. The Vatican claims to have disclosed all of them to the world, but has been accused of holding back part of the third, "Last Secret of Fatima."

believe in miracles: Pew Forum on Religion 2010.

239 *270,000 customers:* Markoff 2007.

According to Clarke's: Clarke 1973 lays out three laws. The first and second are: (1) When a distinguished but elderly scientist states that something is possible, he is almost certainly right. When he states that something is impossible, he is very probably wrong. (2) The only way of discovering the limits of the possible is to venture a little way past them into the impossible.

242 *Sonny Graham received a heart:* Dudley 2008.

Cheryl had been married five times: Wigmore 2008.

244 *Sperm survive the best:* Woods et al. 2004.

ice inside cells is lethal: Mazur, Rall, and Rigopoulos 1981.

still damaging to cells: I use the term *salty* here for convenience, but in reality other solutes beside salt ions are also important.

245 *oocytes and embryos:* Woods et al. 2004.

vitrified kidney: Fahy et al. 2009.

Peter Mazur: Mazur 1988.

246 *"respirator brain":* Towbin 1973.

determination of death: Laureys 2005; President's Council on Bioethics 2008.

247 *"If the brainstem is dead":* Laureys 2005.

vicious cycle continues: President's Council on Bioethics 2008.

248 *discarded information:* Conversely, some of the information in the connectome might be irrelevant to personal identity, because it's just random "noise" created as the brain wired itself up during development.

After mechanical ventilation: Agarwal, Singh, and Gupta 2006.

249 *types of damage present:* Rees 1976; Kalimo et al. 1977.

still intact in the EM images: However, many are depleted of their vesicles containing neurotransmitter. Recall that the strength of a synapse is related to its size, and one measure of size is the number of vesicles. Therefore, information about synaptic strength — information that can be regarded as part of the connectome — may be difficult to recover.

250 *Eric Drexler:* Drexler 1986.

251 *Charles Olson:* Olson 1988.

fixing them in place: Formaldehyde and glutaraldehyde are used to link protein molecules together. An even more toxic fixative, osmium tetroxide, has the dual function of binding together fat molecules and staining the membranes to which they belong.

Figure 53, left: The tissue is embedded in Epon, an epoxy resin, and appears black because of the osmium staining.

252 *Lenin was embalmed:* Modern embalming methods began to develop in the seventeenth and eighteenth centuries. Most notoriously, the eccentric London dentist Martin van

Butchell embalmed his dead wife in 1775 and displayed her in the window of his home office. See Dobson 1953.

15. Save As . . .

255 *"mind uploading":* In his 1955 story "The Tunnel under the World," Frederik Pohl wrote: "Each machine was controlled by a sort of computer which reproduced, in its electronic snarl, the actual memory and mind of a human being. . . . It was only a matter . . . of transferring a man's habit patterns from brain cells to vacuum-tube cells" (Pohl 1956). The first mention in the scientific literature may have been in Martin 1971: "We shall assume that developments in neurobiology, bioengineering and related disciplines . . . will ultimately provide suitable techniques of 'read-out' of the stored information from cryobiologically preserved brains into *n*th generation computers capable of vastly outdoing the dynamic patterning of operation of our cerebral neurones."

257 *requires dying first:* After being resurrected, Jesus is said to have ascended to heaven without dying again. Shoemaker 2002 describes how Christians have argued for millennia over whether the Virgin Mary also entered heaven without dying first. Being carried up to heaven by God is called "Assumption," to distinguish it from the "Ascension" of Jesus, which happened by his own power. In 1950, Pope Pius XII promulgated *Munificentissimus Deus,* which decreed that Mary, "having completed the course of her earthly life, was assumed body and soul into heavenly glory." This dogma recognized the importance of the Assumption but didn't really settle the debates, because its wording was ambiguous. Christians have also long argued over whether the Old Testament figures of Elijah and Enoch were assumed into heaven without dying first.

258 *"brain in a vat":* The story "Where Am I?" in Dennett 1978 is a wonderful example. For an actual attempt to keep an isolated guinea pig brain alive and functioning, see Llinas, Yarom, and Sugimori 1981.

259 *pyramidal tract contains:* Lassek and Rasmussen 1940. For another way of tallying the numbers, let's categorize the neurons of a nervous system by their connection to the outside world. Sensory neurons convert external stimuli into neural signals. For example, the photoreceptors of the retina produce electrical signals when stimulated by light. Motor neurons make synapses onto muscles and convert neural signals into movements. The remainder are called interneurons, because they are interposed between sensory and motor neurons. In the *C. elegans* nervous system, sensory neurons, motor neurons, and interneurons are found in comparable numbers. But sensory and motor neurons make up a vanishingly small fraction of our nervous system. Saying that a neuron is an interneuron is no great distinction, because almost all are. Very few of the neurons in our brains "talk" with the outside world. They mostly talk with each other.
running on a gigantic computer: Bostrom 2003; Lloyd 2006.
Alan Turing: Turing 1950.

260 *successful example of AI:* There are some slight differences in Turing's original setup of the test. The interested reader should consult Turing's paper, which is very readable.
a proper Turing test: Natalie Zemon Davis has argued that Guerre's wife knew very well that the new Guerre was fake, but fell in love and conspired with him (Davis 1983, 1988). But no historians question that some of Guerre's sisters and friends were genuinely fooled.

262 *The more accurate the simulation:* Then again, self-models are often not very accurate. Researchers have shown that most people have inflated opinions of their own abilities. This is called the Lake Wobegon Effect, after the humorist Garrison Keillor's fictional town in which "all the women are strong, all the men are good-looking, and all the children are above average."

263 *Markram was one of the first:* He also showed that the strength of a cortical synapse can fluctuate from spike to spike. In collaboration with theoretical colleagues, he introduced mathematical models describing this phenomenon, known as short-term synaptic plasticity.

simulation of a cat brain: Ananthanarayanan et al. 2009.

264 *"Cat Fight Brews Over Cat Brain":* Adee 2009 also prints the full text of the letter.

266 *neurons of the same type:* For example, when neuroscientists inject electrical current into an inhibitory neuron of the neocortex, it can generate spikes for a long time without faltering (Connors and Gutnick 1990). But when they stimulate a pyramidal neuron, it slows down after the first few spikes, as if it were becoming "fatigued."

Once all neuron types: It will also be necessary to classify synapses into types. Here I've taken the view that neuron types already include all information about synapse types. According to Dale's Principle, a neuron secretes the same neurotransmitter (or set of neurotransmitters) at all of the synapses it makes onto other neurons. That's why all the outgoing synapses of a pyramidal neuron secrete glutamate. There are many variants of glutamate receptor molecules. The particular variant that occurs at a synapse may be a property of the neuron type of the receiving neuron. In other words, the type of a synapse may be determined by the types of the neurons that it connects. If this turns out not to be true, then connectomes will have to include separate information about synapse types as well as neuron types.

millions of ion channels: This numerical estimate is courtesy of Michael Hausser and Arnd Roth. The multicompartmental models are based on the aggregate behaviors of large populations of channels. This has some similarity to the way in which pollsters keep track of the percentage of voters who support a candidate. Each compartment represents some part of the neuronal membrane. It contains multiple populations of ion channels, one population for each channel type. Therefore, if a neuron is divided into one hundred compartments, and there are ten types of ion channel, then the model contains a thousand variables for specifying the states of the ion channels. That may sound like a lot of variables, but it's still much less than the total number of ion channels in the neuron.

multicompartmental model neurons: Multicompartmental models are essential when different parts of a neuron function independently. The dendrites of a single starburst amacrine cell of the retina, for example, detect multiple directions of visual motion and send different signals to other neurons (Euler, Detwiler, and Denk 2002).

Peters' Rule: This was first stated in its general form by Braitenberg and Schüz 1998, and named in honor of Alan Peters for formulating a specific case of the rule.

267 *more difficult for* C. elegans: Lockery and Goodman 2009.

The only information unique: More realistically, the properties of each neuron type might vary slightly across normal people. These variations might be predictable from their genomes. If so, we'd have to say "You are your connectome plus models of neuron types plus your genome." But again, a genome contains much less information than a connectome, so "You are your connectome" would still be a good approximation.

about one hundred types: White et al. 1986.

269 *diffusion of neurotransmitter:* Electronic circuits sometimes behave differently from their simulations, in which components can interact only if they are connected by wires. A real circuit can contain interactions mediated by "thin air" rather than wires. For example, one wire can set up an electric field that is felt by a nearby wire, a phenomenon known as "stray capacitance" that is analogous to extrasynaptic interactions in the brain. This type of deviation from the model can be extremely difficult to identify and troubleshoot.

almost beyond imagining: If you're up to the mind-bending task of thinking about such a simulation, you can consult Tipler 1994, which proves that it should be possible in this universe.

all the positions and velocities: I'm avoiding the issue of whether quantum physics is important for the functioning of the brain. Tegmark 2000 provides some insight into the subject.

271 *Ralph Merkle:* Merkle 1992. Some of the earliest writings about connectomics were penned by proponents of cryonics and uploading, although the term *connectome* was not coined until later. In his 1989 technical report, "The Large Scale Analysis of Neural Structures," Ralph Merkle reviewed the state of the art in serial electron microscopy. He knew that the *C. elegans* connectome had been mapped, and speculated about scaling up to the human brain.

REFERENCES

Abeles, M. 1982. *Local cortical circuits: An electrophysiological study.* Berlin: Springer.

Aboitiz, F., A. B. Scheibel, R. S. Fisher, and E. Zaidel. 1992. Fiber composition of the human corpus callosum. *Brain Research,* 598 (1–2): 143–153.

Abraham, Carolyn. 2002. *Possessing genius: The bizarre odyssey of Einstein's brain.* New York: St. Martin's Press.

Adee, S. 2009. Cat fight brews over cat brain. IEEE Spectrum Tech Talk Blog. Nov. 23.

Agarwal, R., N. Singh, and D. Gupta. 2006. Is the patient brain-dead? *Emergency Medicine Journal,* 23 (1): e05.

Albertson, D. G., and J. N. Thomson. 1976. The pharynx of *Caenorhabditis elegans. Philosophical Transactions of the Royal Society of London, Series B, Biological Sciences,* 275 (938): 299–325.

Amari, S. I. 1972. Learning patterns and pattern sequences by self-organizing nets of threshold elements. *IEEE Transactions on Computers,* 100 (21): 1197–1206.

Amit, D. J. 1989. *Modeling brain function.* Cambridge, Eng.: Cambridge University Press.

Amit, D. J., H. Gutfreund, and H. Sompolinsky. 1985. Spin-glass models of neural networks. *Physical Review A,* 32 (2): 1007.

Amunts, K., G. Schlaug, L. Jäncke, H. Steinmetz, A. Schleicher, A. Dabringhaus, and K. Zilles. 1997. Motor cortex and hand motor skills: Structural compliance in the human brain. *Human Brain Mapping,* 5 (3): 206–215.

Ananthanarayanan, R., S. K. Esser, H. D. Simon, and D. S. Modha. 2009. The cat is out of the bag: Cortical simulations with 10^9 neurons, 10^{13} synapses. In *Proceedings of the Conference on High Performance Computing Networking, Storage, and Analysis,* p. 63. ACM.

Andersen, B. B., L. Korbo, and B. Pakkenberg. 1992. A quantitative study of the human cerebellum with unbiased stereological techniques. *Journal of Comparative Neurology,* 326 (4): 549.

Antonini, A., and M. P. Stryker. 1993. Development of individual geniculocortical arbors in cat striate cortex and effects of binocular impulse blockade. *Journal of Neuroscience,* 13 (8): 3549.

———. 1996. Plasticity of geniculocortical afferents following brief or prolonged monocular occlusion in the cat. *Journal of Comparative Neurology,* 369 (1): 64–82.

Azevedo, F. A., L. R. Carvalho, L. T. Grinberg, J. M. Farfel, R. E. Ferretti, R. E. Leite, F. W. Jacob, R. Lent, and S. Herculano-Houzel. 2009. Equal numbers of neuronal and nonneuronal cells make the human brain an isometrically scaled-up primate brain. *Journal of Comparative Neurology,* 513 (5): 532–541.

Bagwell, C. E. 2005. "Respectful image": Revenge of the barber surgeon. *Annals of Surgery,* 241 (6): 872.

Bailey, A., A. Le Couteur, I. Gottesman, P. Bolton, E. Simonoff, E. Yuzda, and M. Rutter. 1995. Autism as a strongly genetic disorder: Evidence from a British twin study. *Psychological Medicine,* 25 (1): 63–77.

Bailey, P., and G. von Bonin. 1951. *The isocortex of man.* Urbana, Ill.: University of Illinois Press.

Bamman, M. M., B. R. Newcomer, D. E. Larson-Meyer, R. L. Weinsier, and G. R. Hunter. 2000. Evaluation of the strength-size relationship in vivo using various muscle size indices. *Medicine and Science in Sports and Exercise,* 32 (7): 1307.

Barlow, H. B. 1972. Single units and sensation: A neuron doctrine for perceptual psychology. *Perception,* 1 (4): 371–394.

Basser, L. S. 1962, Hemiplegia of early onset and the faculty of speech with special reference to the effects of hemispherectomy. *Brain,* 85: 427–460.

Baum, K. M., and E. F. Walker. 1995. Childhood behavioral precursors of adult symptom dimensions in schizophrenia. *Schizophrenia Research,* 16 (2): 111–120.

Bear, M. F., B. W. Connors, and M. Paradiso. 2007. *Neuroscience: Exploring the brain,* 3rd ed. Baltimore: Lippincott, Williams, and Wilkins.

Beard, M. 2008. *The fires of Vesuvius: Pompeii lost and found.* Cambridge, Mass.: Harvard University Press.

Bechtel, W. 2006. *Discovering cell mechanisms: The creation of modern cell biology.* Cambridge, Eng.: Cambridge University Press.

Benes, F. M., M. Turtle, Y. Khan, and P. Farol. 1994. Myelination of a key relay zone in the hippocampal formation occurs in the human brain during childhood, adolescence, and adulthood. *Archives of General Psychiatry,* 51 (6): 477–484.

Bernal, B., and N. Altman. 2010. The connectivity of the superior longitudinal fasciculus: A tractography DTI study. *Magnetic Resonance Imaging,* 28 (2): 217–225.

Bertone, T., and G. De Carli. 2008. *The last secret of Fatima.* New York: Doubleday.

BGW. 2002. Graduate student in peril: A first person account of schizophrenia. *Schizophrenia Bulletin,* 28 (4): 745–755.

Bhardwaj, R. D., M. A. Curtis, K. L. Spalding, B. A. Buchholz, D. Fink, T. Björk-Eriksson, C. Nordborg, F. H. Gage, H. Druid, P. S. Eriksson, et al. 2006. Neocortical neurogenesis in humans is restricted to development. *Proceedings of the National Academy of Sciences*, 103 (33): 12564.

Bi, G., and M. Poo. 1998. Synaptic modifications in cultured hippocampal neurons: Dependence on spike timing, synaptic strength, and postsynaptic cell type. *Journal of Neuroscience*, 18 (24): 10464.

Blakeslee, Sandra. 2000. A decade of discovery yields a shock about the brain. *New York Times*, Jan. 4.

Boatman, D., J. Freeman, E. Vining, M. Pulsifer, D. Miglioretti, R. Minahan, B. Carson, J. Brandt, and G. McKhann. 1999. Language recovery after left hemispherectomy in children with late-onset seizures. *Annals of Neurology*, 46 (4): 579–586.

Bock, D. D., W. C. A. Lee, A. M. Kerlin, M. L. Andermann, G. Hood, A. W. Wetzel, S. Yurgenson, E. R. Soucy, H. S. Kim, and R. C. Reid. 2011. Network anatomy and in vivo physiology of visual cortical neurons. *Nature*, 471 (7337): 177–182.

Bock, O., and G. Kommerell. 1986. Visual localization after strabismus surgery is compatible with the "outflow" theory. *Vision Research*, 26 (11): 1825.

Bosch, F., and L. Rosich. 2008. The contributions of Paul Ehrlich to pharmacology: A tribute on the occasion of the centenary of his Nobel prize. *Pharmacology*, 82 (3): 171–179.

Bosl, W., A. Tierney, H. Tager-Flusberg, and C. Nelson. 2011. EEG complexity as a biomarker for autism spectrum disorder risk. *BMC Medicine*, 9: 18.

Bostrom, N. 2003. Are you living in a computer simulation? *Philosophical Quarterly*, 53 (211): 243–255.

Bouchard, T. J., Jr., D. T. Lykken, M. McGue, N. L. Segal, and A. Tellegen. 1990. Sources of human psychological differences: The Minnesota Study of Twins Reared Apart. *Science*, 250: 223–228.

Boyke, J., J. Driemeyer, C. Gaser, C. Buchel, and A. May. 2008. Training-induced brain structure changes in the elderly. *Journal of Neuroscience*, 28 (28): 7031.

Bradley, G. D. 1920. *The story of the Pony Express*, 4th ed. Chicago: McClurg.

Braitenberg, V., and A. Schüz. 1998. *Cortex: Statistics and geometry of neuronal connectivity*. Berlin: Springer.

Briggman, K. L., M. Helmstaedter, and W. Denk. 2011. Wiring specificity in the direction-selectivity circuit of the retina. *Nature*, 471 (7337): 183–188.

Brodmann, K. 1909. *Vergleichende Lokalisationslehre der Großhirnrinde in ihren Prinzipien dargestellt auf Grund des Zellenbaues*. Leipzig: Barth. English trans. available as Garey, L. J. 2006. *Brodmann's localisation in the cerebral cortex: The principles of comparative localisation in the cerebral cortex based on cytoarchitectonics*. New York: Springer.

Bruer, J. T. 1999. *The myth of the first three years: A new understanding of early brain development and lifelong learning*. New York: Free Press.

Brundin, P., J. Karlsson, M. Emgård, G. S. Kaminski Schierle, O. Hansson, Å Petersén, and R. F. Castilho. 2000. Improving the survival of grafted dopaminergic neurons: A review over current approaches. *Cell Transplantation*, 9 (2): 179–196.

Bullock, T. H., M. V. L. Bennett, D. Johnston, R. Josephson, E. Marder, and R. D. Fields. 2005. The neuron doctrine, redux. *Science*, 310 (5749): 791.

Buonomano, D. V., and M. M. Merzenich. 1998. Cortical plasticity: From synapses to maps. *Annual Review of Neuroscience*, 21 (1): 149–186.

Burrell, Brian. 2004. *Postcards from the brain museum: The improbable search for meaning in the matter of famous minds*. New York: Broadway Books.

Buss, R. R., W. Sun, and R. W. Oppenheim. 2006. Adaptive roles of programmed cell death during nervous system development. *Annual Review of Neuroscience*, 29: 1.

Cardno, A. G., and I. I. Gottesman. 2000. Twin studies of schizophrenia: From bow-and-arrow concordances to star wars Mx and functional genomics. *American Journal of Medical Genetics, C, Seminars in Medical Genetics*, 97 (1): 12–17.

Carmichael, S. T. 2006. Cellular and molecular mechanisms of neural repair after stroke: Making waves. *Annals of Neurology*, 59 (5): 735–742.

Carper, R. A., P. Moses, Z. D. Tigue, and E. Courchesne. 2002. Cerebral lobes in autism: Early hyperplasia and abnormal age effects. *Neuroimage*, 16 (4): 1038–1051.

Catani, M., and D. H. ffytche. 2005. The rises and falls of disconnection syndromes. *Brain*, 128: 2224–2239.

Chadwick, J. 1960. *The decipherment of Linear B*. Cambridge, Eng.: Cambridge University Press.

Chalfie, M., J. E. Sulston, J. G. White, E. Southgate, J. N. Thomson, and S. Brenner. 1985. The neural circuit for touch sensitivity in *Caenorhabditis elegans*. *Journal of Neuroscience*, 5 (4): 956.

Changeux, Jean-Pierre. 1985. *Neuronal man: The biology of mind*. New York: Pantheon.

Chen, B. L., D. H. Hall, and D. B. Chklovskii. 2006. Wiring optimization can relate neuronal structure and function. *Proceedings of the National Academy of Sciences*, 103 (12): 4723.

Chklovskii, D. B., and A. A. Koulakov. 2004. Maps in the brain: What can we learn from them? *Annual Review of Neuroscience*, 27: 369–392.

Clarke, Arthur C. 1973. *Profiles of the future: An inquiry into the limits of the possible*, rev. ed. New York: Harper & Row.

CMS Collaboration. 2008. The CMS experiment at the CERN LHC. *Journal of Instrumentation*, 3: S08004.

Cohen, L. G., P. Celnik, A. Pascual-Leone, B. Corwell, L. Faiz, J. Dambrosia, M. Honda, N. Sadato, C. Gerloff, M. D. Catalá, et al. 1997. Functional relevance of cross-modal plasticity in blind humans. *Nature*, 389 (6647): 180–183.

Coleman, M. 2005. Axon degeneration mechanisms: Commonality amid diversity. *Nature Reviews Neuroscience*, 6 (11): 889–898.

Conel, J. L. 1939–1967. *Postnatal development of the human cerebral cortex.* 8 vols. Cambridge, Mass.: Harvard University Press.

Conforti, L., R. Adalbert, and M. P. Coleman. 2007. Neuronal death: Where does the end begin? *Trends in Neurosciences,* 30 (4): 159–166.

Connors, B. W., and M. J. Gutnick. 1990. Intrinsic firing patterns of diverse neocortical neurons. *Trends in Neurosciences,* 13 (3): 99–104.

Corkin, S. 2002. What's new with the amnesic patient HM? *Nature Reviews Neuroscience,* 3 (2): 153–160.

Courchesne, E., and K. Pierce. 2005. Why the frontal cortex in autism might be talking only to itself: Local over-connectivity but long-distance disconnection. *Current Opinion in Neurobiology,* 15 (2): 225–230.

Courchesne, E., K. Pierce, C. M. Schumann, E. Redcay, J. A. Buckwalter, D. P. Kennedy, and J. Morgan. 2007. Mapping early brain development in autism. *Neuron,* 56 (2): 399–413.

Cowan, W. M., J. W. Fawcett, D. D. O'Leary, and B. B. Stanfield. 1984. Regressive events in neurogenesis. *Science,* 225 (4668): 1258.

Cramer, S. C. 2008. Repairing the human brain after stroke: I. Mechanisms of spontaneous recovery. *Annals of Neurology,* 63 (3): 272–287.

Davis, Kenneth C. 2005. *Don't know much about mythology: Everything you need to know about the greatest stories in human history but never learned.* New York: HarperCollins.

Davis, N. Z. 1983. *The Return of Martin Guerre.* Cambridge, Mass.: Harvard University Press.

———. 1988. On the lame. *American Historical Review,* 93 (3): 572–603.

DeFelipe, J. 2010. *Cajal's butterflies of the soul: Science and art.* New York: Oxford University Press.

DeFelipe, J., and E. G. Jones. 1988. *Cajal on the cerebral cortex.* New York: Oxford University Press.

Denk, W., and H. Horstmann. 2004. Serial block-face scanning electron microscopy to reconstruct three-dimensional tissue nanostructure. *PLoS Biology,* 2 (11): e329.

Dennett, Daniel Clement. 1978. *Brainstorms: Philosophical essays on mind and psychology.* Montgomery, Vt.: Bradford Books.

Desimone, R., T. D. Albright, C. G. Gross, and C. Bruce. 1984. Stimulus-selective properties of inferior temporal neurons in the macaque. *Journal of Neuroscience,* 4 (8): 2051.

Devlin, K. 2010. *The unfinished game: Pascal, Fermat, and the seventeenth-century letter that made the world modern.* New York: Basic Books.

Dobell, C. C. 1960. *Antony van Leeuwenhoek and his "little animals."* New York: Dover.

Dobson, J. 1953. Some eighteenth century experiments in embalming. *Journal of the History of Medicine and Allied Sciences,* 8 (Oct.): 431.

Doupe, A. J., and P. K. Kuhl. 1999. Birdsong and human speech: Common themes and mechanisms. *Annual Review of Neuroscience,* 22 (1): 567–631.

Draaisma, D. 2000. *Metaphors of memory: A history of ideas about the mind.* Cambridge, Eng.: Cambridge University Press.

Draganski, B., C. Gaser, V. Busch, G. Schuierer, U. Bogdahn, and A. May. 2004. Neuroplasticity: Changes in grey matter induced by training. *Nature,* 427 (6972): 311–312.

Draganski, B., C. Gaser, G. Kempermann, H. G. Kuhn, J. Winkler, C. Buchel, and A. May. 2006. Temporal and spatial dynamics of brain structure changes during extensive learning. *Journal of Neuroscience,* 26 (23): 6314.

Drexler, K. E. 1986. *Engines of creation: The coming era of nanotechnology.* New York: Anchor.

Dronkers, N. F., O. Plaisant, M. T. Iba-Zizen, and E. A. Cabanis. 2007. Paul Broca's historic cases: High resolution MR imaging of the brains of Leborgne and Lelong. *Brain,* 130 (5): 1432.

Dudley, R. 2008. Suicide claims two men who shared one heart. islandpacket.com. Apr. 5.

Eccles, J. C. 1965. Possible ways in which synaptic mechanisms participate in learning, remembering and forgetting. *Anatomy of Memory,* 1: 12–87.

———. 1976. From electrical to chemical transmission in the central nervous system. *Notes and Records of the Royal Society of London,* 30 (2): 219.

Eccles, J. C., P. Fatt, and K. Koketsu. 1954. Cholinergic and inhibitory synapses in a pathway from motor-axon collaterals to motoneurones. *Journal of Physiology,* 126 (3): 524.

Edelman, Gerald M. 1987. *Neural Darwinism: The theory of neuronal group selection.* New York: Basic Books.

Eichenbaum, H. 2000. A cortical-hippocampal system for declarative memory. *Nature Reviews Neuroscience,* 1 (1): 41–50.

Elbert, T., and B. Rockstroh. 2004. Reorganization of human cerebral cortex: The range of changes following use and injury. *Neuroscientist,* 10 (2): 129.

Elbert, T., C. Pantev, C. Wienbruch, B. Rockstroh, and E. Taub. 1995. Increased cortical representation of the fingers of the left hand in string players. *Science,* 270 (5234): 305.

Eling, P., ed. 1994. *Reader in the history of aphasia: From Franz Gall to Norman Geschwind.* Amsterdam: John Benjamins.

Epsztein, J., M. Brecht, and A. K. Lee. 2011. Intracellular determinants of hippocampal CA1 place and silent cell activity in a novel environment. *Neuron,* 70 (1): 109–120.

Euler, T., P. B. Detwiler, and W. Denk. 2002. Directionally selective calcium signals in dendrites of starburst amacrine cells. *Nature,* 418 (6900): 845–852.

Fahy, G. M., B. Wowk, R. Pagotan, A. Chang, J. Phan, B. Thomson, and L. Phan. 2009. Physical and biological aspects of renal vitrification. *Organogenesis,* 5 (3): 167.

Fee, M. S., A. A. Kozhevnikov, and R. H. Hahnloser. 2004. Neural mechanisms of vocal sequence generation in the songbird. *Annals of the New York Academy of Sciences*, 1016: 153.

Fehér, O., H. Wang, S. Saar, P. P. Mitra, and O. Tchernichovski. 2009. De novo establishment of wild-type song culture in the zebra finch. *Nature*, 459 (7246): 564–568.

Felleman, D. J., and D. C. Van Essen. 1991. Distributed hierarchical processing in the primate cerebral cortex. *Cerebral Cortex*, 1 (1): 1.

Fiala, J. C. 2005. Reconstruct: A free editor for serial section microscopy. *Journal of Microscopy*, 218 (1): 52–61.

Fields, R. D. 2009. *The other brain*. New York: Simon & Schuster.

Fiete, I. R., W. Senn, C. Z. H. Wang, and R. H. R. Hahnloser. 2010. Spike-time-dependent plasticity and heterosynaptic competition organize networks to produce long scale-free sequences of neural activity. *Neuron*, 65 (4): 563–576.

Finger, S. 2005. *Minds behind the brain: A history of the pioneers and their discoveries*. New York: Oxford University Press.

Finger, S., and M. P. Hustwit. 2003. Five early accounts of phantom limb in context: Pare, Descartes, Lemos, Bell, and Mitchell. *Neurosurgery*, 52 (3): 675.

Flatt, A. E. 2005. Webbed fingers. *Proceedings (Baylor University Medical Center)*, 18 (1): 26.

Flechsig, P. 1901. Developmental (myelogenetic) localisation of the cerebral cortex in the human subject. *Lancet*, 158 (4077): 1027–1030.

Fombonne, E. 2009. Epidemiology of pervasive developmental disorders. *Pediatric Research*, 65 (6): 591–598.

Ford, B. J. 1985. *Single lens: The story of the simple microscope*. New York: Harper & Row.

Friederici, A. D. 2009. Pathways to language: Fiber tracts in the human brain. *Trends in Cognitive Sciences*, 13 (4): 175–181.

Friston, K. J. 1998. The disconnection hypothesis. *Schizophrenia Research*, 30 (2): 115–125.

Frith, U. 1993. Autism. *Scientific American*, 268 (6): 108–114.

———. 2008. *Autism: A very short introduction*. New York: Oxford University Press.

Frost, D. O., D. Boire, G. Gingras, and M. Ptito. 2000. Surgically created neural pathways mediate visual pattern discrimination. *Proceedings of the National Academy of Sciences*, 97 (20): 11068.

Fukuchi-Shimogori, T., and E. A. Grove. 2001. Neocortex patterning by the secreted signaling molecule FGF8. *Science*, 294 (5544): 1071.

Fukunaga, T., M. Miyatani, M. Tachi, M. Kouzaki, Y. Kawakami, and H. Kanehisa. 2001. Muscle volume is a major determinant of joint torque in humans. *Acta Physiologica Scandinavica*, 172 (4): 249–255.

Gall, F. J. 1835. *On the functions of the brain and of each of its parts: With observations on the possibility of determining the instincts, propensities, and talents, or the moral and intellectual dispositions of men and animals, by the con-figuration of the brain and head.* Trans. W. Lewis. Boston: Marsh, Capen & Lyon.

Galton, F. 1889. On head growth in students at the University of Cambridge. *Journal of Anthropological Institute of Great Britain and Ireland,* 18: 155–156.

———. 1908. *Memories of my life.* London: Methuen.

Gaser, C., and G. Schlaug. 2003. Brain structures differ between musicians and non-musicians. *Journal of Neuroscience,* 23 (27): 9240.

Gelbard-Sagiv, H., R. Mukamel, M. Harel, R. Malach, and I. Fried. 2008. Internally generated reactivation of single neurons in human hippocampus during free recall. *Science,* 322 (5898): 96.

Geschwind, D. H., and P. Levitt. 2007. Autism spectrum disorders: Developmental disconnection syndromes. *Current Opinion in Neurobiology,* 17 (1): 103–111.

Geschwind, N. 1965a. Disconnexion syndromes in animals and man, i. *Brain,* 88 (2): 237–294.

———. 1965b. Disconnexion syndromes in animals and man, ii. *Brain,* 88 (3): 585–644.

Gilbert, M., R. Busund, A. Skagseth, P. Å. Nilsen, and J. P. Solbø. 2000. Resuscitation from accidental hypothermia of 137°C with circulatory arrest. *Lancet,* 355 (9201): 375–376.

Glahn, D. C., J. D. Ragland, A. Abramoff, J. Barrett, A. R. Laird, C. E. Bearden, and D. I. Velligan. 2005. Beyond hypofrontality: A quantitative meta-analysis of functional neuroimaging studies of working memory in schizophrenia. *Human Brain Mapping,* 25 (1): 60–69.

Gould, E., A. J. Reeves, M. S. A. Graziano, and C. G. Gross. 1999. Neurogenesis in the neocortex of adult primates. *Science,* 286 (5439): 548–552.

Greenough, W. T., J. E. Black, and C. S. Wallace. 1987. Experience and brain development. *Child Development,* 58 (3): 539–559.

Gross, C. G. 2000. Neurogenesis in the adult brain: Death of a dogma. *Nature Reviews Neuroscience,* 1 (1): 67–73.

———. 2002. Genealogy of the "grandmother cell." *Neuroscientist,* 8 (5): 512.

Guerrini, R., and E. Parrini. 2010. Neuronal migration disorders. *Neurobiology of Disease,* 38 (2): 154–166.

Guillery, R. W. 2005. Observations of synaptic structures: Origins of the neuron doctrine and its current status. *Philosophical Transactions B,* 360 (1458): 1281.

Hahnloser, R. H. R., A. A. Kozhevnikov, and M. S. Fee. 2002. An ultra-sparse code underlies the generation of neural sequences in a songbird. *Nature,* 419 (6902): 65–70.

Hajszan, T., N. J. MacLusky, and C. Leranth. 2005. Short-term treatment with the antidepressant fluoxetine triggers pyramidal dendritic spine synapse formation in rat hippocampus. *European Journal of Neuroscience,* 21 (5): 1299–1303.

Hall, D. H., and Z. F. Altun. 2008. *C. elegans* atlas. Cold Spring Harbor, N.Y.: Cold Spring Harbor Laboratory Press.

Hall, D. H., and R. L. Russell. 1991. The posterior nervous system of the nematode *Caenorhabditis elegans:* Serial reconstruction of identified neurons and complete pattern of synaptic interactions. *Journal of Neuroscience,* 11 (1): 1.

Hallmayer, J., S. Cleveland, A. Torres, J. Phillips, B. Cohen, T. Torigoe, J. Miller, A. Fedele, J. Collins, K. Smith, et al. 2011. Genetic heritability and shared environmental factors among twin pairs with autism. *Archives of General Psychiatry.* doi: 10.1001/archgenpsychiatry.2011.76

Harris, J. C. 2003. Pinel orders the chains removed from the insane at Bicêtre. *Archives of General Psychiatry,* 60 (5): 442.

Häusser, M., N. Spruston, and G. J. Stuart. 2000. Diversity and dynamics of dendritic signaling. *Science,* 290 (5492): 739.

He, H. Y., W. Hodos, and E. M. Quinlan. 2006. Visual deprivation reactivates rapid ocular dominance plasticity in adult visual cortex. *Journal of Neuroscience,* 26 (11): 2951–2955.

Hebb, D. O. 1949. *The organization of behavior: A neuropsychological theory.* New York: Wiley.

Hell, S. W. 2007. Far-field optical nanoscopy. *Science,* 316 (5828): 1153–1158.

Helmstaedter, M., K. L. Briggman, and W. Denk. 2008. 3D structural imaging of the brain with photons and electrons. *Current Opinion in Neurobiology,* 18 (6): 633–641.

———. High-accuracy neurite reconstruction for high-throughput neuroanatomy. *Nature Neuroscience,* 14 (8): 1081–1088.

Hickok, G., and D. Poeppel. 2007. The cortical organization of speech processing. *Nature Reviews Neuroscience,* 8 (5): 393–402.

Hopfield, J. J. 1982. Neural networks and physical systems with emergent collective computational abilities. *Proceedings of the National Academy of Sciences,* 79 (8): 2554.

Hopfield, J. J., and D. W. Tank. 1986. Computing with neural circuits: A model. *Science,* 233 (4764): 625.

Howland, D. 1996. *Borders of Chinese civilization: Geography and history at empire's end.* Durham, N.C.: Duke University Press.

Hutchinson, S., L. H. L. Lee, N. Gaab, and G. Schlaug. 2003. Cerebellar volume of musicians. *Cerebral Cortex,* 13 (9): 943.

Huttenlocher, P. R. 1990. Morphometric study of human cerebral cortex development. *Neuropsychologia,* 28 (6): 517.

Huttenlocher, P. R., and A. S. Dabholkar. 1997. Regional differences in synaptogenesis in human cerebral cortex. *Journal of Comparative Neurology,* 387 (2): 167–178.

Huttenlocher, P. R., C. de Courten, L. J. Garey, and H. Van der Loos. 1982. Synaptogenesis in human visual cortex-evidence for synapse elimination during normal development. *Neuroscience Letters,* 33 (3): 247–252.

Illingworth, C. M. 1974. Trapped fingers and amputated finger tips in children. *Journal of Pediatric Surgery*, 9 (6): 853–858.

Jain, V., H. S. Seung, and S. C. Turaga. 2010. Machines that learn to segment images: A crucial technology for connectomics. *Current Opinion in Neurobiology*, 20 (5): 653–666.

Jarvis, E. D., O. Güntürkün, L. Bruce, A. Csillag, H. Karten, W. Kuenzel, L. Medina, G. Paxinos, D. J. Perkel, T. Shimizu, et al. 2005. Avian brains and a new understanding of vertebrate brain evolution. *Nature Reviews Neuroscience*, 6 (2): 151–159.

Johansen-Berg, H., and M. F. S. Rushworth. 2009. Using diffusion imaging to study human connectional anatomy. *Annual Review of Neuroscience*, 32: 75–94.

Johnson, L., and S. Baldyga. 2009. *Frozen: My journey into the world of cryonics, deception, and death*. New York: Vanguard.

Jones, P. E. 1995. Contradictions and unanswered questions in the Genie case: A fresh look at the linguistic evidence. *Language and Communication*, 15 (3): 261–280.

Jun, J. K., and D. Z. Jin. 2007. Development of neural circuitry for precise temporal sequences through spontaneous activity, axon remodeling, and synaptic plasticity. *PLoS One*, 2 (8): e273.

Jung, R. E., and R. J. Haier. 2007. The parieto-frontal integration theory (P-FIT) of intelligence: Converging neuroimaging evidence. *Behavioral and Brain Sciences*, 30 (2): 135–154.

Kahn, D. 1967. *The codebreakers: The story of secret writing*. New York: Macmillan.

Kaiser, M. D., C. M. Hudac, S. Shultz, S. M. Lee, C. Cheung, A. M. Berken, B. Deen, N. B. Pitskel, D. R. Sugrue, A. C. Voos, et al. 2010. Neural signatures of autism. *Proceedings of the National Academy of Sciences*, 107 (49): 21223–21228.

Kalil, R. E., and G. E. Schneider. 1975. Abnormal synaptic connections of the optic tract in the thalamus after midbrain lesions in newborn hamsters. *Brain Research*, 100 (3): 690.

Kalimo, H., J. H. Garcia, Y. Kamijyo, J. Tanaka, and B. F. Trump. 1977. The ultrastructure of "brain death." II. Electron microscopy of feline cortex after complete ischemia. *Virchows Archiv B Cell Pathology*, 25 (1): 207–220.

Kanner, L. 1943. Autistic disturbances of affective contact. *Nervous Child*, 2 (2): 217–230.

Karten, H. J. 1997. Evolutionary developmental biology meets the brain: The origins of mammalian cortex. *Proceedings of the National Academy of Sciences*, 94 (7): 2800–2804.

Keith, A. 1927. The brain of Anatole France. *British Medical Journal*, 2 (3491): 1048.

Keller, M. B., J. P. McCullough, D. N. Klein, B. Arnow, D. L. Dunner, A. J. Gelenberg, J. C. Markowitz, C. B. Nemeroff, J. M. Russell, M. E. Thase, et al. 2000. A comparison of nefazodone, the cognitive behavioral-analysis system of psychotherapy, and their combination for the treatment of chronic depression. *New England Journal of Medicine*, 342 (20): 1462–1470.

Keller, S. S., T. Crow, A. Foundas, K. Amunts, and N. Roberts. 2009. Broca's area: Nomenclature, anatomy, typology, and asymmetry. *Brain and Language,* 109 (1): 29–48.

Kelly, Kevin. 1994. *Out of control: The rise of neo-biological civilization.* Reading, Mass.: Addison-Wesley.

Kempermann, G. 2002. Why new neurons? Possible functions for adult hippocampal neurogenesis. *Journal of Neuroscience,* 22 (3): 635.

Kessler, R. C., O. Demler, R. G. Frank, M. Olfson, H. A. Pincus, E. E. Walters, P. Wang, K. B. Wells, and A. M. Zaslavsky. 2005. Prevalence and treatment of mental disorders, 1990 to 2003. *New England Journal of Medicine,* 352 (24): 2515.

Kim, I. J., Y. Zhang, M. Yamagata, M. Meister, and J. R. Sanes. 2008. Molecular identification of a retinal cell type that responds to upward motion. *Nature,* 452 (7186): 478–482.

Knott, G., H. Marchman, D. Wall, and B. Lich. 2008. Serial section scanning electron microscopy of adult brain tissue using focused ion beam milling. *Journal of Neuroscience,* 28 (12): 2959.

Knudsen, E. I., and P. F. Knudsen. 1990. Sensitive and critical periods for visual calibration of sound localization by barn owls. *Journal of Neuroscience,* 10 (1): 222.

Kola, I., and J. Landis. 2004. Can the pharmaceutical industry reduce attrition rates? *Nature Reviews Drug Discovery,* 3 (8): 711–716.

Kolb, B., and R. Gibb. 2007. Brain plasticity and recovery from early cortical injury. *Developmental Psychobiology,* 49 (2): 107–118.

Kolodkin, A. L., and M. Tessier-Lavigne. 2011. Mechanisms and molecules of neuronal wiring: A primer. *Cold Spring Harbor Perspectives in Biology,* 3: a001727.

Kolodzey, J. 1981. Cray-1 computer technology. *IEEE Transactions on Components, Hybrids, and Manufacturing Technology,* 4 (2): 181–186.

Kornack, D. R., and P. Rakic. 1999. Continuation of neurogenesis in the hippocampus of the adult macaque monkey. *Proceedings of the National Academy of Sciences,* 96 (10): 5768.

———. 2001. Cell proliferation without neurogenesis in adult primate neocortex. *Science,* 294 (5549): 2127.

Kostovic, I., and P. Rakic. 1980. Cytology and time of origin of interstitial neurons in the white matter in infant and adult human and monkey telencephalon. *Journal of Neurocytology,* 9 (2): 219.

Kozel, F. A., K. A. Johnson, Q. Mu, E. L. Grenesko, S. J. Laken, and M. S. George. 2005. Detecting deception using functional magnetic resonance imaging. *Biological Psychiatry,* 58 (8): 605–613.

Kubicki, M., H. Park, C. F. Westin, P. G. Nestor, R. V. Mulkern, S. E. Maier, M. Niznikiewicz, E. E. Connor, J. J. Levitt, M. Frumin, et al. 2005. DTI and MTR abnormalities in schizophrenia: Analysis of white matter integrity. *Neuroimage,* 26 (4): 1109–1118.

Kullmann, D. M. 2010. Neurological channelopathies. *Annual Review of Neuroscience*, 33: 151–172.

Lander, E. S. 2011. Initial impact of the sequencing of the human genome. *Nature*, 470 (7333): 187–197.

Langleben, D. D., L. Schroeder, J. A. Maldjian, R. C. Gur, S. McDonald, J. D. Ragland, C. P. O'Brien, and A. R. Childress. 2002. Brain activity during simulated deception: An event-related functional magnetic resonance study. *Neuroimage*, 15 (3): 727–732.

Lashley, K. S. 1929. *Brain mechanisms and intelligence: A quantitative study of injuries to the brain.* Chicago: University of Chicago Press.

Lashley, K. S., and G. Clark. 1946. The cytoarchitecture of the cerebral cortex of *Ateles:* A critical examination of architectonic studies. *Journal of Comparative Neurology*, 85 (2): 223–305.

Lassek, A. M., and G. L. Rasmussen. 1940. A comparative fiber and numerical analysis of the pyramidal tract. *Journal of Comparative Neurology*, 72 (2): 417–428.

Laureys, S. 2005. Death, unconsciousness, and the brain. *Nature Reviews Neuroscience*, 6 (11): 899–909.

Lederberg, J., and A. T. McCray. 2001. 'Ome sweet 'omics: A genealogical treasury of words. *Scientist*, 15 (7): 8.

Leeuwenhoek, A. van. 1674. More Observations from Mr. Leewenhook, in a Letter of Sept. 7. 1674. Sent to the publisher. *Philosophical Transactions*, 9 (108): 178–182.

Legrand, N., A. Ploss, R. Balling, P. D. Becker, C. Borsotti, N. Brezillon, and J. Debarry. 2009. Humanized mice for modeling human infectious disease: Challenges, progress, and outlook. *Cell Host and Microbe*, 6 (1): 5–9.

Leroi, A. 2006. What makes us human? *Telegraph*, Aug. 1.

Leucht, S., C. Corves, D. Arbter, R. R. Engel, C. Li, and J. M. Davis. 2009. Second-generation versus first-generation antipsychotic drugs for schizophrenia: A meta-analysis. *Lancet*, 373 (9657): 31–41.

Lewis, D. A., and P. Levitt. 2002. Schizophrenia as a disorder of neurodevelopment. *Annual Review of Neuroscience*, 25: 409.

Lichtman, J. W., and H. Colman. 2000. Synapse elimination review and indelible memory. *Neuron*, 25: 269–278.

Lichtman, J. W., J. R. Sanes, and J. Livet. A technicolour approach to the connectome. *Nature Reviews Neuroscience*, 9 (6): 417–422.

Lieberman, P. 2002. On the nature and evolution of the neural bases of human language. *American Journal of Physical Anthropology*, 119 (S35): 36–62.

Lindbeck, A. 1995. The prize in economic science in memory of Alfred Nobel. *Journal of Economic Literature*, 23 (1): 37–56.

Linkenhoker, B. A., and E. I. Knudsen. 2002. Incremental training increases the plasticity of the auditory space map in adult barn owls. *Nature*, 419 (6904): 293–296.

Lipton, P. 1999. Ischemic cell death in brain neurons. *Physiological Reviews*, 79 (4): 1431–1568.

Livet, J., T. A. Weissman, H. Kang, J. Lu, R. A. Bennis, J. R. Sanes, and J. W. Lichtman. 2007. Transgenic strategies for combinatorial expression of fluorescent proteins in the nervous system. *Nature*, 450 (7166): 56–62.

Lledo, P. M., M. Alonso, and M. S. Grubb. 2006. Adult neurogenesis and functional plasticity in neuronal circuits. *Nature Reviews Neuroscience*, 7 (3): 179–193.

Llinas, R., Y. Yarom, and M. Sugimori. 1981. Isolated mammalian brain in vitro: New technique for analysis of electrical activity of neuronal circuit function. *Federation Proceedings*, 40: 2240.

Lloyd, Seth. 2006. *Programming the universe: A quantum computer scientist takes on the cosmos.* New York: Knopf.

Lockery, S. R., and M. B. Goodman. 2009. The quest for action potentials in *C. elegans* neurons hits a plateau. *Nature Neuroscience*, 12 (4): 377–378.

Lopez-Munoz, F., and C. Alamo. 2009. Monoaminergic neurotransmission: The history of the discovery of antidepressants from 1950s until today. *Current Pharmaceutical Design*, 15 (14): 1563–1586.

Lotze, M., H. Flor, W. Grodd, W. Larbig, and N. Birbaumer. 2001. Phantom movements and pain: An fMRI study in upper limb amputees. *Brain*, 124 (11): 2268.

Machin, Geoffrey. 2009. Non-identical monozygotic twins, intermediate twin types, zygosity testing, and the non-random nature of monozygotic twinning: A review. *American Journal of Medical Genetics, C, Seminars in Medical Genetics*, 151C (2): 110–127.

Maguire, E. A., D. G. Gadian, I. S. Johnsrude, C. D. Good, J. Ashburner, R. S. J. Frackowiak, and C. D. Frith. 2000. Navigation-related structural change in the hippocampi of taxi drivers. *Proceedings of the National Academy of Sciences*, 97 (8): 4398.

Markoff, J. 2007. Already, Apple sells refurbished iPhones. *New York Times*, Aug. 22.

Markou, A., C. Chiamulera, M. A. Geyer, M. Tricklebank, and T. Steckler. 2008. Removing obstacles in neuroscience drug discovery: The future path for animal models. *Neuropsychopharmacology*, 34 (1): 74–89.

Markram, H., J. Lubke, M. Frotscher, and B. Sakmann. 1997. Regulation of synaptic efficacy by coincidence of postsynaptic APs and EPSPs. *Science*, 275 (5297): 213.

Marr, D. 1971. Simple memory: A theory for archicortex. *Philosophical Transactions of the Royal Society of London. Series B, Biological Sciences*, 262 (841): 23–81.

Martin, G. M. 1971. Brief proposal on immortality: An interim solution. *Perspectives in Biology and Medicine*, 14 (2): 339.

Mashour, G. A., E. E. Walker, and R. L. Martuza. 2005. Psychosurgery: Past, present, and future. *Brain Research Reviews*, 48 (3): 409–419.

Masland, R. H. 2001. Neuronal diversity in the retina. *Current Opinion in Neurobiology*, 11 (4): 431–436.

Mathern, G. W. 2010. Cerebral hemispherectomy. *Neurology*, 75 (18): 1578.

Maughan, R. J., J. S. Watson, and J. Weir. 1983. Strength and cross-sectional area of human skeletal muscle. *Journal of Physiology*, 338 (1): 37.

Matzelle, T. R., H. Gnaegi, A. Ricker, and R. Reichelt. 2003. Characterization of the cutting edge of glass and diamond knives for ultramicrotomy by scanning force microscopy using cantilevers with a defined tip geometry: Part 2. *Journal of Microscopy*, 209 (2): 113–117.

Mazur, P. 1988. Stopping biological time. *Annals of the New York Academy of Sciences*, 541 (1): 514–531.

Mazur, P., W. F. Rall, and N. Rigopoulos. 1981. Relative contributions of the fraction of unfrozen water and of salt concentration to the survival of slowly frozen human erythrocytes. *Biophysical Journal*, 36 (3): 653–675.

McClelland, J. L., and D. E. Rumelhart. 1981. An interactive activation model of context effects in letter perception: I. An account of basic findings. *Psychological Review*, 88 (5): 375.

McDaniel, M. A. 2005. Big-brained people are smarter: A meta-analysis of the relationship between in vivo brain volume and intelligence. *Intelligence*, 33 (4): 337–346.

Mechelli, A., J. T. Crinion, U. Noppeney, J. O'Doherty, J. Ashburner, R. S. Frackowiak, and C. J. Price. 2004. Neurolinguistics: Structural plasticity in the bilingual brain. *Nature*, 431 (7010): 757.

Mendez, I., A. Viñuela, A. Astradsson, K. Mukhida, P. Hallett, H. Robertson, T. Tierney, R. Holness, A. Dagher, J. Q. Trojanowski, et al. 2008. Dopamine neurons implanted into people with Parkinson's disease survive without pathology for 14 years. *Nature Medicine*, 14 (5): 507–509.

Merkle, R. C. 1992. The technical feasibility of cryonics. *Medical Hypotheses*, 39 (1): 6–16.

Mesulam, M. M. 1998. From sensation to cognition. *Brain*, 121 (6): 1013.

Meyer, M. P., and S. J. Smith. 2006. Evidence from in vivo imaging that synaptogenesis guides the growth and branching of axonal arbors by two distinct mechanisms. *Journal of Neuroscience*, 26 (13): 3604.

Mezard, M., G. Parisi, and M. A. Virasoro. 1987. *Spin glass theory and beyond*. Singapore: World Scientific.

Micale, M. S. 1985. The Salpêtrière in the age of Charcot: An institutional perspective on medical history in the late nineteenth century. *Journal of Contemporary History*, 20 (4): 703–731.

Middleton, F. A., and P. L. Strick. 2000. Basal ganglia output and cognition: Evidence from anatomical, behavioral, and clinical studies. *Brain and Cognition*, 42 (2): 183–200.

Miles, M., and D. Beer. 1996. Pakistan's microcephalic chuas of Shah Daulah: Cursed, clamped, or cherished. *History of Psychiatry*, 7 (28, pt. 4): 571.

Miller, K. D. 1996. Synaptic economics: Competition and cooperation in correlation-based synaptic plasticity. *Neuron*, 17: 371–374.

Minsky, M. 2006. *The emotion machine.* New York: Simon & Schuster.

Mochida, G. H., and C. A. Walsh. 2001. Molecular genetics of human microcephaly. *Current Opinion in Neurology*, 14 (2): 151.

———. 2004. Genetic basis of developmental malformations of the cerebral cortex. *Archives of Neurology*, 61 (5): 637.

Mochizuki, H. 2009. Parkin gene therapy. *Parkinsonism and Related Disorders*, 15: S43–S45.

Mohr, J. P. 1976. Broca's area and Broca's aphasia. In Haiganoosh Whitaker and Harry A. Whitaker, eds., *Studies in neurolinguistics*, vol. 1, *Perspectives in neurolinguistics and psycholinguistics*, pp. 201–235. New York: Academic Press.

Mooney, R., and J. F. Prather. 2005. The HVC microcircuit: The synaptic basis for interactions between song motor and vocal plasticity pathways. *Journal of Neuroscience*, 25 (8): 1952–1964.

Morgan, S., P. Grootendorst, J. Lexchin, C. Cunningham, and D. Greyson. 2011. The cost of drug development: A systematic review. *Health Policy*, 100 (1): 4–17.

Murphy, B. P., Y. C. Chung, T. W. Park, and P. D. McGorry. 2006. Pharmacological treatment of primary negative symptoms in schizophrenia: A systematic review. *Schizophrenia Research*, 88 (1–3): 5–25.

Murphy, T. H., and D. Corbett. 2009. Plasticity during stroke recovery: From synapse to behaviour. *Nature Reviews Neuroscience*, 10 (12): 861–872.

Myrdal, G. 1997. The Nobel Prize in economic science. *Challenge*, 20 (1): 50–52.

Nehlig, A. 2010. Is caffeine a cognitive enhancer? *Journal of Alzheimer's Disease*, 20 (Supp): S85–S94.

Nelson, S. B., K. Sugino, and C. M. Hempel. 2006. The problem of neuronal cell types: A physiological genomics approach. *Trends in Neurosciences*, 29 (6): 339–345.

Nestler, E. J., and S. E. Hyman. 2010. Animal models of neuropsychiatric disorders. *Nature Neuroscience*, 13 (10): 1161–1169.

Newhouse, P. A., A. Potter, and A. Singh. 2004. Effects of nicotinic stimulation on cognitive performance. *Current Opinion in Pharmacology*, 4 (1): 36–46.

Nicolelis, M. 2007. Living with Ghostly Limbs. *Scientific American Mind*, 18 (6): 52–59.

O'Connell, M. R. 1997. *Blaise Pascal: Reasons of the heart.* Grand Rapids, Mich.: Wm. B. Eerdmans.

Oddo, S., A. Caccamo, J. D. Shepherd, M. P. Murphy, T. E. Golde, R. Kayed, R. Metherate, M. P. Mattson, Y. Akbari, and F. M. LaFerla. 2003. Triple-transgenic model of Alzheimer's disease with plaques and tangles: Intracellular Aβ and synaptic dysfunction. *Neuron*, 39 (3): 409–421.

Olanow, C. W., C. G. Goetz, J. H. Kordower, A. J. Stoessl, V. Sossi, M. F. Brin, K. M.

Shannon, G. M. Nauert, D. P. Perl, J. Godbold, et al. 2003. A double-blind controlled trial of bilateral fetal nigral transplantation in Parkinson's disease. *Annals of Neurology*, 54 (3): 403–414.

Olshausen, B. A., C. H. Anderson, and D. C. Van Essen. 1993. A neurobiological model of visual attention and invariant pattern recognition based on dynamic routing of information. *Journal of Neuroscience*, 13 (11): 4700–4719.

Olson, C. B. 1988. A possible cure for death. *Medical Hypotheses*, 26 (1): 77–84.

Pakkenberg, B., and H. J. Gundersen. 1997. Neocortical neuron number in humans: Effect of sex and age. *Journal of Comparative Neurology*, 384 (2): 312.

Passingham, R. E., K. E. Stephan, and R. Kötter. 2002. The anatomical basis of functional localization in the cortex. *Nature Reviews Neuroscience*, 3 (8): 606–616.

Paterniti, Michael. 2000. *Driving Mr. Albert: A trip across America with Einstein's brain.* New York: Dial.

Paul, L. K., W. S. Brown, R. Adolphs, J. M. Tyszka, L. J. Richards, P. Mukherjee, and E. H. Sherr. 2007. Agenesis of the corpus callosum: Genetic, developmental and functional aspects of connectivity. *Nature Reviews Neuroscience*, 8 (4): 287–299.

Pearson, K. 1906. On the relationship of intelligence to size and shape of head, and to other physical and mental characters. *Biometrika*, 5 (1–2): 105.

———. 1924. *The life, letters and labours of Francis Galton.* Vol. 2, *Researches of middle life.* London: Cambridge University Press.

Peck, D. T. 1998. Anatomy of an historical fantasy: The Ponce de León fountain of youth legend. *Revista de Historia de América*, 123: 63–87.

Penfield, W., and E. Boldrey. 1937. Somatic motor and sensory representation in the cerebral cortex of man as studied by electrical stimulation. *Brain*, 60 (4): 389.

Penfield, W., and T. Rasmussen. 1952. *The cerebral cortex of man.* New York: Macmillan.

Petrie, W. M. F. 1883. *The pyramids and temples of Gizeh.* London: Field & Tuer.

Pew Forum on Religion. 2010. Religion among the millennials. Technical report, Pew Research Center, Feb.

Plum, F. 1972. Prospects for research on schizophrenia, 3. Neurophysiology. Neuropathological findings. *Neurosciences Research Program Bulletin*, 10 (4): 384.

Poeppel, D., and G. Hickok. 2004. Towards a new functional anatomy of language. *Cognition*, 92 (1–2): 1–12.

Pohl, Frederik. 1956. *Alternating currents.* New York: Ballantine.

Porter, K. R., and J. Blum. 1953. A study in microtomy for electron microscopy. *Anatomical Record*, 117 (4): 685–709.

President's Council on Bioethics. 2008. Controversies in the determination of death. Washington, D.C.

Purves, D. 1990. *Body and brain: A trophic theory of neural connections.* Cambridge, Mass.: Harvard University Press.

Purves, D., L. E. White, and D. R. Riddle. 1996. Is neural development Darwinian? *Trends in Neurosciences,* 19 (11): 460–464.

Purves, Dale, and Jeff W. Lichtman. 1985. *Principles of neural development.* Sunderland, Mass.: Sinauer Associates.

Quiroga, R. Q., L. Reddy, G. Kreiman, C. Koch, and I. Fried. 2005. Invariant visual representation by single neurons in the human brain. *Nature,* 435 (7045): 1102–1107.

Rakic, P. 1985. Limits of neurogenesis in primates. *Science,* 227 (4690): 1054.

Rakic, P., J. P. Bourgeois, M. F. Eckenhoff, N. Zecevic, and P. S. Goldman-Rakic. 1986. Concurrent overproduction of synapses in diverse regions of the primate cerebral cortex. *Science,* 232 (4747): 232–235.

Ramachandran, V. S., and S. Blakeslee. 1999. *Phantoms in the brain: Probing the mysteries of the human mind.* New York: Harper Perennial.

Ramachandran, V. S., M. Stewart, and D. C. Rogers-Ramachandran. 1992. Perceptual correlates of massive cortical reorganization. *Neuroreport,* 3 (7): 583.

Ramón y Cajal, Santiago. 1921. Textura de la corteza visual del gato. *Archivos de Neurobiología,* 2: 338–362. Trans. in DeFelipe and Jones 1988.

———. 1989. *Recollections of my life.* Cambridge, Mass.: MIT Press.

Rapoport, J. L., A. M. Addington, S. Frangou, and MRC Psych. 2005. The neurodevelopmental model of schizophrenia: Update 2005. *Molecular Psychiatry,* 10 (5): 434–449.

Rasmussen, T., and B. Milner. 1977. The role of early left-brain injury in determining lateralization of cerebral speech functions. *Annals of the New York Academy of Sciences,* 299: 355–369.

Redcay, E., and E. Courchesne. 2005. When is the brain enlarged in autism? A meta-analysis of all brain size reports. *Biological Psychiatry,* 58 (1): 1–9.

Rees, S. 1976. A quantitative electron microscopic study of the ageing human cerebral cortex. *Acta Neuropathologica,* 36 (4): 347–362.

Reilly, K. T., and A. Sirigu. 2008. The motor cortex and its role in phantom limb phenomena. *Neuroscientist,* 14 (2): 195.

Rilling, J. K. 2008. Neuroscientific approaches and applications within anthropology. *American Journal of Physical Anthropology,* 137 (S47): 2–32.

Rilling, J. K., and T. R. Insel. 1998. Evolution of the cerebellum in primates: Differences in relative volume among monkeys, apes and humans. *Brain, Behavior, and Evolution,* 52 (6): 308.

———. 1999. The primate neocortex in comparative perspective using magnetic resonance imaging. *Journal of Human Evolution,* 37 (2): 191–223.

Robinson, A. 2002. *Lost languages: The enigma of the world's undeciphered scripts.* New York: McGraw-Hill.

Rosenzweig, M. R. 1996. Aspects of the search for neural mechanisms of memory. *Annual Review of Psychology,* 47 (1): 1–32.

Ruestow, E. G. 1983. Images and ideas: Leeuwenhoek's perception of the spermatozoa. *Journal of the History of Biology,* 16 (2): 185–224.

————. 1996. *The microscope in the Dutch Republic: The shaping of discovery.* New York: Cambridge University Press.

Rumelhart, David E., and James L. McClelland. 1986. *Parallel distributed processing: Explorations in the microstructure of cognition.* Cambridge, Mass.: MIT Press.

Russell, R. M. 1978. The CRAY-1 computer system. *Communications of the ACM,* 21 (1): 63–72.

Ruthazer, E.S., J. Li, and H. T. Cline. 2006. Stabilization of axon branch dynamics by synaptic maturation. *Journal of Neuroscience,* 26 (13): 3594.

Rymer, R. 1994. *Genie: A scientific tragedy.* New York: HarperPerennial.

Sadato, N., A. Pascual-Leone, J. Grafman, V. Ibanez, M. P. Deiber, G. Dold, and M. Hallett. 1996. Activation of the primary visual cortex by Braille reading in blind subjects. *Nature,* 380 (6574): 526–528.

Sahay, A., and R. Hen. 2007. Adult hippocampal neurogenesis in depression. *Nature Neuroscience,* 10 (9): 1110–1115.

Sale, A., J. F. M. Vetencourt, P. Medini, M. C. Cenni, L. Baroncelli, R. De Pasquale, and L. Maffei. 2007. Environmental enrichment in adulthood promotes amblyopia recovery through a reduction of intracortical inhibition. *Nature Neuroscience,* 10 (6): 679–681.

Schildkraut, J. J. 1965. The catecholamine hypothesis of affective disorders: A review of supporting evidence. *American Journal of Psychiatry,* 122 (5): 509–522.

Schiller, F. 1963. Leborgne — in memoriam. *Medical History,* 7 (1): 79.

————. 1992. *Paul Broca: Founder of French anthropology, explorer of the brain.* New York: Oxford University Press.

Schmahmann, J. D. 2010. The role of the cerebellum in cognition and emotion: Personal reflections since 1982 on the dysmetria of thought hypothesis, and its historical evolution from theory to therapy. *Neuropsychology Review,* 20 (3): 236–260.

Schneider, G. E. 1973. Early lesions of superior colliculus: Factors affecting the formation of abnormal retinal projections. *Brain, Behavior and Evolution,* 8 (1): 73.

————. 1979. Is it really better to have your brain lesion early? A revision of the "Kennard principle." *Neuropsychologia,* 17 (6): 557.

Schüz, A., D. Chaimow, D. Liewald, and M. Dortenman. 2006. Quantitative aspects of corticocortical connections: A tracer study in the mouse. *Cerebral Cortex,* 16 (10): 1474.

Selfridge, O. G. Pattern recognition and modern computers. 1955. In *Proceedings of the March 1–3, 1955, Western Joint Computer Conference,* pp. 91–93. ACM.

Seligman, M. 2011. *Flourish: A visionary new understanding of happiness and well-being.* New York: Free Press.

Selkoe, D. J. 2002. Alzheimer's disease is a synaptic failure. *Science,* 298 (5594): 789.

Seung, H. S. 2009. Reading the book of memory: Sparse sampling versus dense mapping of connectomes. *Neuron*, 62 (1): 17–29.

Shen, W. W. 1999. A history of antipsychotic drug development. *Comprehensive Psychiatry*, 40 (6): 407–414.

Shendure, J., R. D. Mitra, C. Varma, and G. M. Church. 2004. Advanced sequencing technologies: Methods and goals. *Nature Reviews Genetics*, 5 (5): 335–344.

Sherrington, C. S. 1924. Problems of muscular receptivity. *Nature*, 113 (2851): 894–894.

Shoemaker, Stephen J. 2002. *Ancient traditions of the Virgin Mary's dormition and assumption.* Oxford: Oxford University Press.

Sizer, Nelson. 1888. *Forty years in phrenology.* New York: Fowler & Wells.

Soldner, F., D. Hockemeyer, C. Beard, Q. Gao, G. W. Bell, E. G. Cook, G. Hargus, A. Blak, O. Cooper, M. Mitalipova, et al. 2009. Parkinson's disease patient-derived induced pluripotent stem cells free of viral reprogramming factors. *Cell*, 136 (5): 964–977.

Song, S., P. J. Sjostrom, M. Reigl, S. Nelson, and D. B. Chklovskii. 2005. Highly nonrandom features of synaptic connectivity in local cortical circuits. *PLoS Biol*, 3 (3): e68.

Sporns, O., J. P. Changeux, D. Purves, L. White, and D. Riddle. 1997. Variation and selection in neural function: Authors' reply. *Trends in Neurosciences*, 20 (7): 291–293.

Sporns, O., G. Tononi, and R. Kotter. 2005. The human connectome: A structural description of the human brain. *PLoS Comput Biol*, 1 (4): e42.

Spurzheim, J. G. 1833. *A view of the elementary principles of education: Founded on the study of the nature of man.* Boston: Marsh, Capen & Lyon.

Steen, R. G., C. Mull, R. Mcclure, R. M. Hamer, and J. A. Lieberman. 2006. Brain volume in first-episode schizophrenia: Systematic review and meta-analysis of magnetic resonance imaging studies. *British Journal of Psychiatry*, 188 (6): 510.

Steffenburg, S., C. Gillberg, L. Hellgren, L. Andersson, I. C. Gillberg, G. Jakobsson, and M. Bohman. 1989. A twin study of autism in Denmark, Finland, Iceland, Norway, and Sweden. *Journal of Child Psychology and Psychiatry*, 30 (3): 405–416.

Stent, G. S. 1973. A physiological mechanism for Hebb's postulate of learning. *Proceedings of the National Academy of Sciences*, 70 (4): 997.

Sterr, A., M. M. Müller, T. Elbert, B. Rockstroh, C. Pantev, and E. Taub. 1998. Perceptual correlates of changes in cortical representation of fingers in blind multifinger Braille readers. *Journal of Neuroscience*, 18 (11): 4417.

Stevens, C. F. 1998. Neuronal diversity: Too many cell types for comfort? *Current Biology*, 8 (20): R708-R710.

Stratton, G. M. 1897a. Vision without inversion of the retinal image: Part 1. *Psychological Review*, 4 (4): 341–360.

———. 1897b. Vision without inversion of the retinal image: Part 2. *Psychological Review,* 4 (5): 463–481.

Strebhardt, K., and A. Ullrich. 2008. Paul Ehrlich's magic bullet concept: 100 years of progress. *Nature Reviews Cancer,* 8 (6): 473–480.

Strick, P. L., R. P. Dum, and J. A. Fiez. 2009. Cerebellum and nonmotor function. *Annual Review of Neuroscience,* 32: 413–434.

Stuart, Greg, Nelson Spruston, and Michael Häusser. 2007. *Dendrites.* Oxford: Oxford University Press.

Sur, M., P. E. Garraghty, and A. W. Roe. 1988. Experimentally induced visual projections into auditory thalamus and cortex. *Science,* 242 (4884): 1437.

Swanson, L. W. 2000. What is the brain? *Trends in Neurosciences,* 23 (11): 519–527.

———. 2012. *Brain architecture: Understanding the basic plan,* 2nd ed. New York: Oxford University Press.

Tang, Y., J. R. Nyengaard, D. M. G. De Groot, and H. J. G. Gundersen. 2001. Total regional and global number of synapses in the human brain neocortex. *Synapse,* 41 (3): 258–273.

Taub, R. 2004. Liver regeneration: From myth to mechanism. *Nature Reviews Molecular Cell Biology,* 5 (10): 836–847.

Tegmark, M. 2000. Why the brain is probably not a quantum computer. *Information Sciences,* 128 (3–4): 155–179.

Tipler, Frank J. 1994. *The physics of immortality: Modern cosmology, God, and the resurrection of the dead.* New York: Doubleday.

Tomasch, J. 1954. Size, distribution, and number of fibres in the human corpus callosum. *Anatomical Record,* 119 (1): 119–135.

Towbin, A. 1973. The respirator brain death syndrome. *Human Pathology,* 4 (4): 583–594.

Treffert, D. A. 2009. The savant syndrome, an extraordinary condition: A synopsis, past, present, future. *Philosophical Transactions of the Royal Society B: Biological Sciences,* 364 (1522): 1351.

Turing, A. M. 1950. Computer machinery and intelligence. *Mind,* 59 (236): 433–460.

Turkheimer, E. 2000. Three laws of behavior genetics and what they mean. *Current Directions in Psychological Science,* 9 (5): 160.

Utter, A. A., and M. A. Basso. 2008. The basal ganglia: An overview of circuits and function. *Neuroscience and Biobehavioral Reviews,* 32 (3): 333–342.

Varshney, L. R., B. L. Chen, E. Paniagua, D. H. Hall, and D. B. Chklovskii. 2011. Structural properties of the *Caenorhabditis elegans* neuronal network. *PLoS Computational Biology,* 7 (2): e1001066.

Vein, A. A., and M. L. C. Maat-Schieman. 2008. Famous Russian brains: Historical attempts to understand intelligence. *Brain,* 131 (2): 583.

Vetencourt, J. F. M., A. Sale, A. Viegi, L. Baroncelli, R. De Pasquale, O. F. O'Leary, E. Castrén, and L. Maffei. 2008. The antidepressant fluoxetine restores plasticity in the adult visual cortex. *Science,* 320 (5874): 385–388.

Vining, E. P. J., J. M. Freeman, D. J. Pillas, S. Uematsu, B. S. Carson, J. Brandt, D.

Boatman, M. B. Pulsifer, and A. Zuckerberg. 1997. Why would you remove half a brain? The outcome of 58 children after hemispherectomy. *Pediatrics*, 100 (2): 163.

Vita, A., L. De Peri, C. Silenzi, and M. Dieci. 2006. Brain morphology in first-episode schizophrenia: A meta-analysis of quantitative magnetic resonance imaging studies. *Schizophrenia Research*, 82 (1): 75–88.

Voigt, J., and H. Pakkenberg. 1983. Brain weight of Danish children: A forensic material. *Acta Anatomica*, 116 (4): 290.

Wang, J. W., D. J. David, J. E. Monckton, F. Battaglia, and R. Hen. 2008. Chronic fluoxetine stimulates maturation and synaptic plasticity of adult-born hippocampal granule cells. *Journal of Neuroscience*, 28 (6): 1374–1384.

West, M. J., and A. P. King. 1990. Mozart's starling. *American Scientist*, 78 (2): 106–114.

White, J. G., E. Southgate, J. N. Thomson, and S. Brenner. 1986. The structure of the nervous system of the nematode *Caenorhabditis elegans*. *Philosophical Transactions of the Royal Society of London. B, Biological Sciences*, 314 (1165): 1.

Wigmore, B. 2008. How tyrant wife 'drove two of her five husbands to suicide' — after one was transplanted with heart of the other. *Daily Mail*, Sept. 1.

Wilkes, A. L., and N. J. Wade. 1997. Bain on neural networks. *Brain and Cognition*, 33 (3): 295–305.

Witelson, S. F., D. L. Kigar, and T. Harvey. 1999. The exceptional brain of Albert Einstein. *Lancet*, 353 (9170): 2149–2153.

Woods, E. J., J. D. Benson, Y. Agca, and J. K. Critser. 2004. Fundamental cryobiology of reproductive cells and tissues. *Cryobiology*, 48 (2): 146–156.

Yamada, M., Y. Mizuno, and H. Mochizuki. 2005. Parkin gene therapy for α-synucleinopathy: A rat model of Parkinson's disease. *Human Gene Therapy*, 16 (2): 262–270.

Yamahachi, H., S. A. Marik, J.N.J. McManus, W. Denk, and C. D. Gilbert. 2009. Rapid axonal sprouting and pruning accompany functional reorganization in primary visual cortex. *Neuron*, 64 (5): 719–729.

Yang, G., F. Pan, and W. B. Gan. 2009. Stably maintained dendritic spines are associated with lifelong memories. *Nature*, 462 (7275): 920–924.

Yates, F. 1966. *The art of memory.* Chicago: University of Chicago Press.

Yuste, Rafael. 2010. *Dendritic spines.* Cambridge, Mass.: MIT Press.

Zhang, R. L., Z. G. Zhang, and M. Chopp. 2005. Neurogenesis in the adult ischemic brain: Generation, migration, survival, and restorative therapy. *Neuroscientist*, 11 (5): 408.

Ziegler, D. A., O. Piquet, D. H. Salat, K. Prince, E. Connally, and S. Corkin. 2010. Cognition in healthy aging is related to regional white matter integrity, but not cortical thickness. *Neurobiology of Aging*, 31 (11): 1912–1926.

Zilles, Karl, and Katrin Amunts. 2010. Centenary of Brodmann's map — conception and fate. *Nature Reviews Neuroscience*, 11 (2): 139–145.

FIGURE CREDITS

Images not credited below are by the author.
Figure 1: Ramón y Cajal 1921; DeFelipe and Jones 1988. Digitized by Javier DeFelipe from the original drawing in the Museo Cajal. Copyright © the heirs of Santiago Ramón y Cajal. **Figure 2:** David H. Hall and Zeynep Altun 2008. Introduction. In Worm Atlas. http://www .wormatlas.org/hermaphrodite/introduction/introframeset.html. **Figure 3:** Copyright © Dmitri Chklovskii, reproduced with permission. *C. elegans* wiring diagram described in Varshney, L. R., B. L. Chen, E. Paniagua, D. H. Hall, and D. B. Chklovskii. Structural properties of the *C. elegans* neuronal network, *PLoS Computational Biology,* 7 (2): e1001066. doi:10.1371/journal .pcbi.1001066 and http://www.hhmi.org/research/groupleaders/chklovskii.html. **Figure 5:** Assembled by Hye-Vin Kim using images from the Benjamin R. Tucker papers, Manuscripts and Archives Division, the New York Public Library, Astor, Lenox and Tilden Foundations. **Figure 6:** Courtesy of David Ziegler and Suzanne Corkin, and part of a study reported in Ziegler et al. 2010. **Figures 7–8:** Rob Duckwall/Dragonfly Media Group. **Figure 9:** Sizer 1888. **Figure 10:** Dronkers, N. F, O. Plaisant, M. T. Iba-Zizen, and E. A. Cabanis. 2007. Paul Broca's historic cases: High resolution MR imaging of the brains of Leborgne and Lelong. *Brain,* 130 (5): 1432–1441. By permission of Oxford University Press. **Figure 11:** Brodmann 1909. **Figure 12:** Penfield and Rasmussen 1954. **Figure 13, left:** David Phillips/Photo Researchers; **right:** Alex K. Shalek, Jacob T. Robinson, and Hongkun Park. **Figure 14:** Constantino Sotelo. See also DeFelipe 2010. **Figure 15:** Ben Mills. **Figure 16, left:** Lawrence Livermore National Laboratory; **right:** copyright © 2009 Andrew Back (Flickr: carrierdetect). **Figure 17:** Albert Lee, Jérôme Epsztein, and Michael Brecht. **Figure 18:** Hye-Vin Kim. **Figure 23:** Yang, G., F. Pan, and W. B. Gan. 2009. Stably maintained dendritic spines are associated with lifelong memories. *Nature,* 462 (7275): 920–924. **Figure 25:** Assembled by Hye-Vin Kim from drawings in Conel 1939–1967. **Figure 26:** Kathy Rockland. **Figure 27:** Hye-Vin Kim. **Figure 28:** Created by Winfried Denk based on an image from Kristen M. Harris, PI, and Josef Spacek. Copyright © SynapseWeb 1999–present. Available at synapses.clm.utexas.edu. **Figure 29:** Courtesy of Kim Peluso, Beaver-Visitec International, Inc.(formerly BD Medical–Ophthalmic Systems). **Figure 30:** Ken Hayworth. **Figure 31:** Richard Schalek. **Figures 32–33:** TEM cross-section of the adult nematode, *C. elegans,* published on www.wormimage.org by David H. Hall, with permission from John White, MRC/ LMB, Cambridge, England. **Figure 34:** Daniel Berger, based on data of Narayanan Kasthuri, Ken Hayworth, Juan Carlos Tapia, Richard Schalek, and Jeff Lichtman. **Figure 35:** Hye-Vin Kim. **Figure 37:** Aleksandar Zlateski. **Figure 38:** Modified from an image provided by Richard Masland. **Figure 39:** Felleman, D. J., and D. C. Van Essen. 1991. Distributed Hierarchical Processing in the Primate Cerebral Cortex. *Cerebral Cortex,* 1 (1): 1–47. By permission of Oxford University Press. **Figure 40, left:** Hye-Vin Kim; **right:** Kathy Rockland. **Figure 41:** Ramón y Cajal 1921; DeFelipe and Jones 1988. Digitized by Javier DeFelipe from the original drawing in

INDEX

Page numbers in italics refer to illustrations.